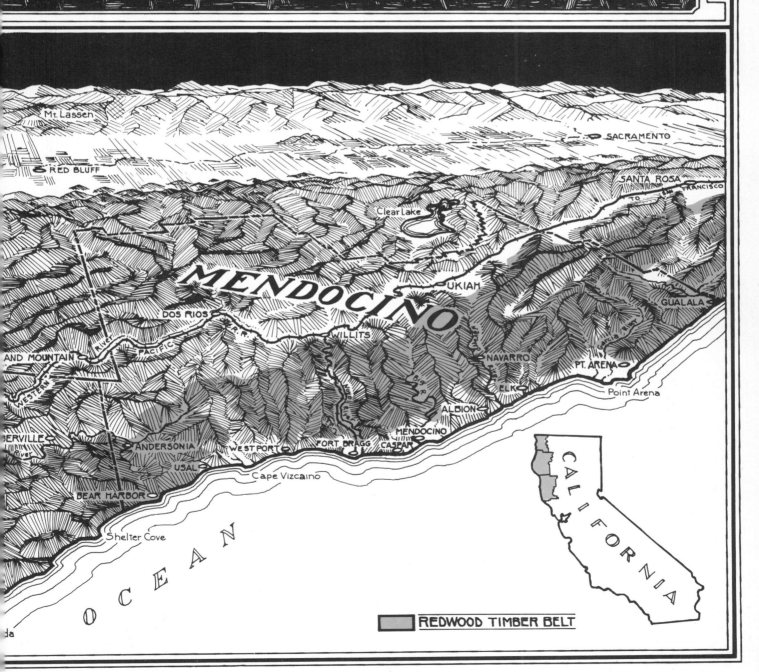

REDWOOD TIMBER BELT

Lynwood Carranco

Golden West Books
San Marino, California · 91108

REDWOOD LUMBER INDUSTRY

Copyright © 1982 by Lynwood Carranco

All Rights Reserved
Published by Golden West Books
San Marino, California 91108 U.S.A.
Library of Congress Catalog Card No. 82-15870
I.S.B.N. 0-87095-084-3

Library of Congress Cataloging in Publication Data

Carranco, Lynwood.
  Redwood Lumber Industry.

    Bibliography: p.
    Includes Index.
    1. Lumber Trade—California—History.
    2. Logging—California—History.
    3. Lumbering—California—History.
    4. Redwood I. Title.
HD9757.C3C35 1982   338.1'7497592   82-15870
ISBN 0-87095-084-3

TITLE PAGE ILLUSTRATION

Sunlight filters through towering redwood trees in Redwood National Park. These trees, which grow 300 feet tall can live 2,000 years, covered some 1.5 million acres of California at the time Columbus discovered America. John Muir once said, "California redwood trees are truly *Avenues of Grandeur.*" — DAVID SWANLUND

Golden West Books
P.O. Box 8136 · San Marino, California · 91108

Dedicated to

EMANUEL FRITZ

The patriarch of California forestry who is recognized as *Mr. Redwood*. He taught at the University of California (Berkeley) from 1919 to 1954, and is internationally recognized through his teaching, publications, and public appearances. He was an early advocate of timber resources renewal and a half-century ago pioneered redwood reforestation.

The chopping boss and a chopper pose to illustrate the height of the undercut of a 19-foot diameter redwood in the Hammond Lumber Company woods in the early 1920's. — OTTO AND MARY EMILY DICK — G.J. SPEIER COLLECTION

Table of Contents

Preface

The beginnings of the redwood lumber industry in northwestern California are rooted in the second half of the nineteenth century, that period in America's history when the concepts of rugged pioneering skill and industrial development walked hand in hand. The story of the evolution of the timber trade, garbed in the complex fabric of corporate dealings, is populated with one example after another of successful exploitation of a major national resource.

Over time, as the mighty monarchs of the California redwood forests crashed to the ground, primitive logging operations carried out by manpower and animal power in the face of well-nigh overwhelming odds gave way to mechanized procedures. Supported by steam power, gasoline power, electrical power, heavy machinery, suitable tools, and technological know-how, man came close to being a match for the topographical and meteorological elements he contended with on the northern Pacific Coast.

After the turn of the century, West Coast lumbering became less and less a local enterprise and more and more a national investment. Outside capital brought outside control, and financial phenomena common to this nation's expansion in the early twentieth century soon appeared in the counties of Humboldt, Mendocino, and Del Norte, the primary California redwood areas.

Today, the ring of axes in the woods, the colorful atmosphere of logging camps, and the exciting aspects of man against Nature that epitomized the first half century or so of American lumbering represent for many no more than a romantic factor in the country's heritage. Moreover, such figures as millions of board feet of lumber and hundreds of thousands of dollars pale to insignificance in an era whose voices speak glibly of "billions."

Nevertheless, the tale of the Northern California timber trade remains eminently worthy of attempts to recount it. Taken together, the past, present, and future of the redwood industry project a microcosm of man's impact on his environment that carries a heartening message for those who are both proud of American enterprise and concerned about the American wilderness: we came, we saw, we conquered, and *we will renew* a magnificent national treasure.

I would like to cite, with sincere thanks, the following good people and institutions for the help that I derived in the compilation of this brief history: Martha Roscoe of the Humboldt County Historical Society; Nannie Escola of the Mendocino County Historical Society; Frances Purser, formerly of the Humboldt State University Library; Robert Krogh of the College of the Redwoods Library; the staff of the Bancroft Library of the University of California at Berkeley; the California Redwood Association in San Francisco; the National Maritime Museum in San Francisco; Lois Bishop of the Louisiana-Pacific Corporation; Paul Evans and James Sharum of the Simpson Timber Company; Stanley Parker and Rodney Wooley of The Pacific Lumber Company; and John Signor who drew the maps.

To my wife, Ruth, for her patience, understanding, and help over the years, I am as always especially grateful.

Lynwood Carranco

September 1982

9

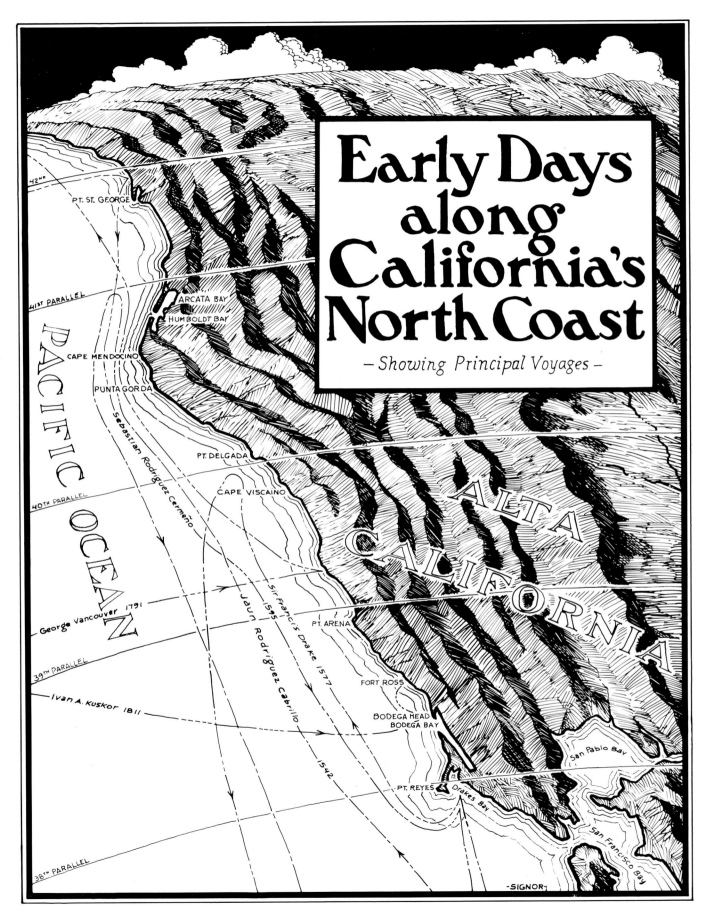

Early Days along California's North Coast

—Showing Principal Voyages—

The Discovery of
Los Palos Colorados

After his conquest of the Aztecs in 1521, Hernán Cortés planned explorations to the northwest along the Pacific coast, confident that treasures commensurate with those of Montezuma were to be found there. By 1522 he had already constructed a shipyard at Zacatula on the Pacific coast of Mexico to serve as a base for sea-going expeditions, only to find his intentions constantly frustrated by lack of local ship-building materials and by the growing distrust of the King of Spain. Not until 1533 was he able to launch the expedition that discovered the lower part of what the Spaniards called "the Californias." Those members of the expedition who survived a hostile reception by the Indians and returned safely to Mexico fed Cortés' hopes with reports of pearls, and two years later he went himself to start a colony at what is now the Bay of La Paz. Between the belligerence of the native Indians and the navigational obstacles to supplying it, however, the little settlement soon collapsed.[1]

Shortly thereafter the first Viceroy of New Spain, Antonio de Mendoza, arrived to take over control of all further official explorations for the legendary wealth waiting to the north. Ignoring Cortés' plans, Mendoza in 1540 sent Coronado, at the head of a bountifully equipped force, to the north and east on a wholly fruitless search for the famous Seven Cities of Cíbola of Indian lore. When Coronado, defeated and disheartened, returned to report the Seven Cities non-existent, the Viceroy turned his attention to still another myth, that of a water route through North America to Asia. To find the Pacific outlet of this fabled strait, he sent Juan Cabrillo north along the coast in an expedition comprised of two small and rickety ships manned by convicts and stocked with a bare minimum of food and water.

Cabrillo sighted Upper California from the sea, and on September 28, 1542 sailed into the harbor of San Diego. He claimed the land for Spain, rested there a few days, and then headed north once more. Not many months later, he died of an infection; but in response to his wishes, the weary crews in their miserable little ships kept beating their way northward. They reached the coast of southern Oregon before the ravages of starvation and scurvy and their inability to find protected harbors forced them to turn back, having in the meantime laid claim to all the unknown land they had sailed by in the name of the Spanish crown.

Thus did Upper or Alta California, so-called at the time to distinguish it from the peninsula of Lower or Baja California, assume the status of

Spanish territory more than two centuries before its colonization began. Since the Cabrillo expedition found no new gateway to Asia, it was rated a failure, and New Spain's interest in the California coast remained limited to an awareness that Spanish galleons sailed past it on their way from Manila to the port of Acapulco, laden with luxurious goods from the Far East purchased with gold and silver from Mexican mines.

By 1593, however, this Asia-Mexico trade had become so heavy that merchants in Spain feared the loss of their markets in Mexico. Since the Spanish mercantile system — like most others of its time — operated on the principle that a colony's economy must take second place to the economy of the mother country, the Spanish monarchy restricted the number of Acapulco-Manila galleons to one per year. Such restraint of trade caused the merchant-adventurer captains of the annual galleons to crowd more and more cargo into space that otherwise would have been used to store food and water, with the inevitable consequence that crews suffered horribly on the long six- to seven-month journey. As a result, the need for a replenishing anchorage at some point on the Pacific coast became apparent, and officials of New Spain attempted to locate such a haven. In 1595 Viceroy Luís de Valasco instructed Sebastián Rodríguez Cermeno, commander of the 200-ton galleon *San Augustin*, to search for a safe California harbor on his return run from Manila. While officers and crew were carrying out this order in a launch they had assembled, catastrophe overtook their anchored galleon. A mammoth southeast gale completely wrecked the ship and demolished her cargo, destroying a whole year's trade!

Such a disaster only confirmed Spanish authorities' already negative attitudes toward the idea of exploring Alta California. Not until 1602 was Sebastián Vizcaíno able to obtain approval of a scheme for finding a safe harbor there — and the investment of considerable sums of his own money may well have been the determining factor in the undertaking. Sailing into what had earlier been named the Bay of San Francisco, Vizcaíno promptly rechristened it Monterey — in honor of the Viceroy of the time — and headed hastily back to Mexico with glowing reports of its suitability, e.g., that it was protected completely from all winds, surrounded by an "abundance of fine timber for ship-building," etc., in order to collect his promised reward, command of the next year's Acapulco-Manila galleon.

Before he could do so, however, a new Viceroy arrived. Distrusting Vizcaíno completely, he revoked the promised command, refused to believe any of the latter's glowing reports of a magnificent harbor at Monterey (correctly enough), and settled down in adamant opposition to the establishment of any port of call in California. Although those who credited Vizcaíno's claims periodically brought up the idea of a coastal colony in Alta California, the leadership of New Spain for the next century and a half viewed that territory as essentially unprofitable and the northward journey as too treacherous and troublesome to be worthwhile. No further effort was made to settle Alta California, and the immense natural wealth of her coastal environment remained unexploited, untamed, and unrealized.

Then, in 1765, the King of Spain appointed José de Gálvez Visitor-General of New Spain. A Visitor-General was a special deputy of the Spanish crown whose powers overlapped and frequently exceeded those of the Viceroy, and Gálvez was the most effective one ever to tackle the New World. At the time, the Spanish king was engaged in expelling all Jesuits from his dominions, and Gálvez's first assignment was to see to the precipitate departure of the Jesuit missionaries who had, over the past seventy years, doggedly and determinedly gone ahead forging a chain of missions in Baja California, under terms of a contract given them by the monarch in 1591. He performed the task promptly, turning the missions over first to the military administration of Captain Gaspar de Portolá and later to the Franciscans.

The most ambitious personal project of José de Gálvez, however, was the extension of the northwestern frontier of New Spain into Alta California, and he went to work on it wholeheartedly and at once. To further his plans, he played heavily and successfully on the long-established fear that some other European power might eventually gain a foothold on the Pacific coast from which to menace Spain. He now added a rumor that Russia was about to do just that to similar tales he himself had been promulgating about both the English and the Dutch, thus considerably strengthening his hand with the Spanish crown, and in 1768 he sailed to the Baja peninsula, there to plan an expedition to colonize Alta California at last.

The following year, a handful of Spanish officers and native Mexican soldiers, under the command of Portolá, and a band of Franciscan missionaries, led by Fray Junípero Serra, set out to establish settlements at both San Diego and Monterey. Serving as chaplain and diarist for the little band was Franciscan Fray Juan Crespí, and his careful records reveal that Portolá, leaving Serra in charge at San Diego, headed north with a small party of men in search of Monterey. They failed to recognize that site when they reached it — a result no doubt of the discrepancy between reality and Vizcaíno's rapturous description of its bay as protected from all winds — and so pushed on to what is now San Francisco Bay. En route, they became the first white men ever to see the giant redwoods near Soquel — more than two centuries after Hernán Cortés had taken the first step in that direction. The entry in Crespí's diary for October 9, 1769 reads:

> ... so we must have traveled but little more than one league, over plains and low hills, well forested with very high trees of a red color, not known to us. They have a very different leaf from cedars, and although the wood resembles cedar somewhat in color, it is very different, and has not the same odor; moreover the wood of the trees we found is very brittle. In this region there is a great abundance of these trees and because none of the expedition recognize them, they are named red wood (*palo colorado*) from their color.[2]

He had penned the first known written reference to the Coast redwoods of the western United States.

Redwood trees have been in existence for millions of years, and some dozen species of them were already native to North America in the Mesozoic Era, when dinosaurs roamed the earth. Only two of those species, however, managed to survive the harsh climatic and environmental upheavals to which they were exposed through ensuing ages. Now commonly referred to as the Sierra redwood and the Coast redwood, they remain today, living near each other in the far western United States.

Botanically classified as *Sequoiadendron giganteum*, the Sierra redwood, a huge cone-bearer or conifer, is the largest known living entity and --at an estimated age of 3,200 years — one of the most ancient. Only one other tree, a bristle-cone pine growing at an elevation of 10,000 feet above California's Death Valley and thought to be 4,600 years old, exceeds it in age.[3]

The tremendous girth of the Sierra redwoods renders logging them both difficult and wasteful, with the fortunate result that they will never be cut down for lumber. Ninety-nine percent of them are safely preserved in national and state parks west of the California-Nevada border along a 260-mile range of the Sierra Nevada Mountains. There they grow, a few to a stand, among other species of trees, at elevations varying from 3,000 to 8,000 feet. Sometimes referred to simply as "the Big Trees," the Sierra redwoods bear tiny, scale-like blue-green leaves and small yellowish cones from one to three inches long. On older trunks, their cinnamon-red bark is deeply furrowed and can be anywhere from one to two feet thick.

Unique as it may be in size and antiquity, the Sierra redwood must give way to its younger and slimmer cousin, the Coast redwood of California and Oregon, when it comes to height. Known to botanists as *Sequoia sempervirens*, this species of redwood is the world's tallest tree, occasionally growing as high as a thirty-storied building. The loftiest example extant is 367.8 feet tall and is situated in the midst of ideal growing conditions on the banks of Redwood Creek in the Tall Tree Grove of the Redwood National Park north of Eureka, in Humboldt County, California. Nearby tower the world's second, third, and sixth tallest trees.

Unlike the Sierra redwoods, the Coast redwoods are highly conducive to logging and have provided commerical lumber since the 1770's. These tall conifer champions of the forests never lose contact with the ocean. Rarely do they grow more than thirty miles inland from the coast or at elevations of more than 2,000 feet; for they thrive on cool sea air and the moisture of Pacific fogs. Most spectacularly visible along a 500-mile strip of coast from Monterey County, California northward to a point just across the Oregon border, their finest stands are to be found on lower slopes and in the deep moist soil of river banks.

The Coast redwoods bear small, pointed green needles, easily distinguishable from the scaly needles of the Sierra redwood. Their cones are tiny, but each one contains hundreds of miniscule seeds. Perhaps the most crucial difference in terms of survival between *Sequoiadendron giganteum* and *Sequoia sempervirens* lies in their reproductive abilities. The Sierra redwood reproduces only from its seeds, while the Coast redwood reproduces both from its seeds and by sprouting anew from its own stump or roots.

Redwood is among the most beautiful and useful of American woods. Its heartwood ranges in color from light cherry to dark mahogany. Its sapwood is pale yellow. Unpainted, it is aesthetically complementary to greenery, as well as to brightly-colored flowers.

Redwood is light in weight, a firm yet soft and easily-worked wood, with a straight fine grain. Perhaps its most important practical characteristic is its durability. Natural chemicals within the heartwood render it highly resistant to attack by termites or decay-causing *fungi*. Thus it has proven an ideal choice for use in gardens and a superlative material for fences, patios, porches, outdoor furniture, pool screens, decks, garden shelters, planters — indeed for any item whose use may bring it into contact with damp ground.

Exposed to such elements as wind and rain, redwood's tendency to warp, split, or swell is minimal. It is also an excellent insulator, because

An aerial view of the Tall Tree Grove on Redwood Creek in the Redwood National Park. The Howard Libbey Tree (367.8 feet), with a dead top, is in the middle of the picture. — DAVID SWANLUND

A beautiful scene in the redwoods on the South Fork of the Eel River in Humboldt County. — DAVID SWANLUND

it retards transmission of both heat and sound. Its cellular structure causes it to hold finishes well, so it is easily painted, stained, bleached, or finished in order to preserve its original hue. Left out of doors unfinished, it will gradually "weather" to a soft and natural silver gray.

Of all soft woods, redwood is the most fire resistant, a quality that has made it highly suitable for outdoor siding and indoor paneling, for doors, window frames, and trim in homes, offices, schools, and churches. Its already-mentioned natural resistance to rot and to wood-eating insects is a boon to industry. For years it has successfully served as water pipe — both buried and above ground — in all kinds of climates. Because it transmits no odor or taste to any liquid contained within it and resists chemical attack, redwood is frequently the material of choice for water storage and for industrial vats and tanks used by wineries, food processors, chemical plants, petroleum refineries, pulp and paper mills, and tanneries.

Paving blocks, railroad ties, furniture cores, suitcase frames — all can be most practicably made of redwood. Moreover, silos, fences, utility sheds, animal shelters, and feeders of redwood provide yeoman service to the American farmer.

El palo colorado was obviously destined to play an indispensable role in the development of California. Indeed, the first redwood lumber trade on the Pacific coast involved building supplies for the mission that Serra had founded at San Diego. In 1776, needing timbers and lumber for that establishment, Don Diego Choquet sailed northward to Monterey on the *San Antonio* (the vessel that had carried the sea-going half of the successful Portola-Serra expedition seven years earlier). Arriving May 21, Choquet took aboard a cargo of pit-sawn timbers and headed homeward on June 30.[4] In 1777, on slopes near the town of San José, Indian workmen used wedges and axes to hand-hew rough timbers for the mission at Santa Clara. By 1812 the Russians — who had finally arrived after all — had built their trading post, Fort Ross, eighteen miles above Bodega Bay, almost entirely out of redwood.[5]

As the population of California began to increase, the demand for lumber rose steadily. Numerous individual inhabitants began providing themselves with small incomes by sawing and selling timber on their own. Both the Monterey and Santa Cruz areas boasted whipsaw mills, and others quickly came into being in what today are the East Bay and the counties of Marin and Santa Clara. The first mills powered by water began operating near Sonoma in 1834. Peter Lassen, a Danish blacksmith, soon built another mill near Mount Hermon in the vicinity of Santa Cruz. He later traded it to Isaac Graham in exchange for 100 mules, and Graham put the mill into operation in 1842.[6]

Thus, by the first half of the 1800's and the start of the *rancho* period, California's mighty northwestern forests had been known to white men for almost a century, and the latter were making increasing use of the towering redwoods. Like the first far-distant rumble of thunder on a hot summer's night, however, the tremendous lumbering potential they represented and awareness of the violence in store for *Sequoia sempervirens* hardly penetrated the consciousness of *los Californios*, preoccupied as they were with more immediate problems and events. The redwoods had been only discovered, not exploited.

Chapter Notes

1. The historical data contained in this chapter covering the period from 1521 to the arrival of Portolá at San Francisco in 1769 is derived from Walter E. Bean, *California, an Interpretive History*, 2nd ed. (New York: McGraw-Hill Book Company, 1973), 13-33 and John Walton Caughey, *California*, 2nd ed. (New Jersey: Prentice-Hall, Inc., 1965), 1-112.
2. Quoted in Herbert Eugene Bolton, ed., *Fray Juan Crespí: Missionary Explorer on the Pacific Coast, 1769-1774* (Berkeley: University of California Press, 1927), 211.
3. Information on redwoods and products of the redwood tree contained in this chapter is derived from personal observations of the author; interviews with lumbermen; a pamphlet entitled *Our Redwood Heritage*, published by the California Redwood Association; a pamphlet entitled *Welcome to Hammond Lumber Company*; and Melville Bell Grosvenor, "World's Tallest Tree Discovered" and Paul Zahl, "Finding the Mt. Everest of All Living Things," *National Geographic* (July 1964): 1-51.
4. Thomas R. Cox, *Mills and Markets: A History of the Pacific Coast Lumber Industry to 1900* (Seattle: University of Washington Press, 1974), 3.
5. *Our Redwood Heritage*.
6. Cox, *Mills and Markets*, 16, 21-22.

Choppers in the Hammond Lumber Company woods in the 1920's stand on staging boards while cutting huge lengths of bark which were shipped out of the county for advertising purposes. — OTTO AND MARY EMILY DICK — G.J. SPEIER COLLECTION

Felling the Giants

The discovery of gold in California in 1848 turned the steady trickle of newcomers into a flood and sent the demand for lumber soaring. Almost simultaneously there appeared in what would become known as "Redwood Country," the three northwestern California counties of Mendocino, Humboldt, and Del Norte, a new breed of man, eager and entrepreneurial, possessed of both capital and business "savvy," and determined to take full advantage of the opportunity to reap a fortune from lumbering.

The manufacture of lumber in Humboldt County began in 1850, but was initially confined to the logging of pine, spruce, and fir trees. The great size and weight of redwoods exceeded the handling and sawing capacities of the primitive facilities on hand at the time. Moreover, the market in San Francisco was not yet specifically demanding redwood to any noticeable extent.[1] A few years later, when redwood logging did begin, machinery in the woods was limited to a hand-operated screw jack that had merely been developed from simpler jacks, rather than especially designed for logging. The saws available were not long enough to fell the giant trees, and old single-bitted axes with helves from thirty-two to thirty-six inches in length had to be used instead. Redwoods were even chopped into log lengths entirely with axes.[2]

In 1953 Peter Rutledge, a former superintendent of the Dolbeer & Carson Lumber Company, who first went to work in the woods in 1889, when he was only fifteen years old, recalled William Carson's description of that earliest redwood cutting:

> The logs were not chopped in a V-shaped cut. The men chopped down a few inches into the tree, and then about two feet away they would chop down another notch into the tree to slab out a piece. The woodsmen continued in this manner until they chopped through the tree. And when they finished, the logs were almost as if they had been cut with a crosscut saw.[3]

In the 1850's many loggers still staked out their own claims and ran their individual outfits. (Those who worked for wages made from $100 to $150 per month.) To bring the logs to the tidewater sloughs, they used two-wheeled trucks pulled by three to four oxen.[4] However, lumber production is by nature a huge undertaking, the kind that requires heavy investment of capital and operations on a massive scale if it is to yield any sizeable profit. The independent logger, therefore, would soon become a figment of the past, as groups of wealthy individuals or organized companies relentlessly acquired most of the timberlands.

SIERRA
REDWOOD

COAST
REDWOOD

Machinery for the early mills had to be shipped by schooner around the Horn, but mill owners had little trouble logging the timber around Humboldt Bay because the land sloped toward the water, which provided a natural log pond for their facilities on the shore. The first railroads in the State of California were those constructed in Humboldt County in 1854 — more than twenty miles of well-graded and substantial roadbed built by loggers —for use in transporting logs to the water's edge. Horses or mules hauled the logs over them on trucks fitted with big axles and large flanged wheels hewn to fit the track or rails, which were usually two parallel rows of straight saplings.[5]

Lumber manufacturing was carried out by means of one or the other of two "plans" or "systems." The more popular and less expensive of these involved production of timber at the point of shipment. Under this method, logs were hauled down from the slopes whole and then sawed into lengths at the mill, which was located on the bay

and accessible to lumber vessels of any draft. The alternative system, employed only when the owners of timber were unable to obtain a mill site with deep-water frontage, called for locating the mill within the timber belt itself, turning logs into lumber there, and then transporting the lumber by rail to deep water for shipment. Another factor to be considered by Humboldt County lumber companies in determining manufacturing procedures was the local topography. In the Mad River area, the land was relatively level. Operations there meant less actual loss of timber than operations on the hillsides. In the Eel River, Freshwater, Ryan Slough, and Jacoby Creek areas, the timber grew on steep hills studded with narrow canyons, and logs had to be sent down by means of chutes. The use of chutes was a far less difficult method of moving logs than hauling them across level country, but it inevitably entailed some loss of timber through breakage.[6]

Any redwood logging operation began with the

cutting of the giant conifers. As indicated earlier, the first redwoods felled were cut completely by axes. The first loggers worked only alongside or close to streams or sloughs, which were deepened to provide small watercourses down which the logs, once afloat, could easily be directed to the mill. Accordingly, these early loggers would cut only a narrow strip of trees on either side of a stream, seldom more than several hundred yards from the water.[7] Another way of moving logs downstream was by means of a sluice or splash dam. Loggers would dam up a creek or river, roll the logs into the stream below the dam, and leave them there pending the arrival of winter floods. The rushing flood waters would break the dam and carry the logs beyond it down to the mill. In Humboldt County, such splash dams were used extensively in Freshwater Creek, Elk River, Mad River, Ryan Slough, and Salmon Creek.

This use of splash dams in Humboldt County was by no means devoid of problems, however. Butt logs — the first log cut from each tree — were often so heavy they would not float, but instead sank in the Elk River. Some sixty or seventy years later, men working with gravel dredges found many of them buried in the riverbed. Still perfectly sound, they were retrieved and used by the split products industry for such items as ties, grape stakes, and posts.[8]

To the south, along the Mendocino coast, streams constituted the sole means of timber transport. Logs were cut in summer and early fall, hauled to the banks of rivers by teams of bulls, and rolled into the beds of streams. There they would lie in immense piles until the first freshets of winter lifted and floated them down to the mills, which were built near the mouths of the rivers. A series of booms adjacent to each mill would reach out to catch and hold them. On occasion, though, an unusually strong freshet would carry away the booms themselves and sweep them, logs and all, out to sea.[9]

The labor of logging was grueling, although its more backbreaking aspects were alleviated somewhat when oxen and horses were brought in to help roll the logs to the stream or river, or to chutes — where the men would dog them together into a string or train of logs and then let the animals drag them down to mill or landing. Nevertheless, the woodsmen covered considerable territory, moving a mile or more upstream every year.[10]

Prior to 1900, monthly wages in the woods were as follows: swampers, $60 to $100; choppers, $65 to $75; sawyers and chain tenders $65 to $100; teamsters, $125 to $180. Board and room was included, and nearly every lumber company maintained three or four cookhouses — one at the mill, one at

A LOGGER'S TOOLS

HANDLE STYLES

REVERSIBLE   HOOP

CLIMAX

SAW TOOTH PATTERNS

REDWOOD KING   REDWOOD FALLING

EUREKA FALLING   EUREKA

WESTERN DOUBLE-BITTED FELLING AXE

"FELLING" CROSSCUT SAW

KEROSENE LUBRICANT FLASK

CAULKED BOOTS

PACIFIC COAST WEDGES

FOR FELLING   FOR LOG MAKING

CROSSCUT BUCKING SAW

-SIGNOR-

THE UNDERCUT 1

TO FALL A TREE...

THE CROSSCUT 2

TIMBER! 3

KEROSENE FLASK

WEDGES

FELLING CROSSCUT SAW

CAULKED BOOTS

STAGING BOARD

DOUBLE BIT AXE

—SIGNOR—

the log dump, and one at each of its camps in the woods.[11] One old-timer who lived in several different camps in the Humboldt County area before the turn of the century recalls that living conditions were good, but he had to furnish his own mattress, blankets, and kerosene.[12]

The early loggers were prone to cut every tree in their path — large or small. Fortunately, however, some of them would leave trees uncut here or there because they did not appear "marketable." These redwoods they left behind have acquired a great degree of importance, for they have revealed that a redwood will grow much more rapidly when deprived of its neighbors than when it is crowded among them. A goodly number of those in question were probably only twenty inches in diameter at breast height when their area was first logged; but now, some eighty to ninety years later, they range from four to six feet across. Moreover, where loggers left a few trees standing with spaces between them, new redwoods from sprouts and

seedlings have grown up to fill the gaps and present some fine examples of second growth.[13]

By 1878 the method of felling a Coast redwood had improved considerably as a result of experience, as well as of newer and more suitable equipment for the job. Long enough saws had become available, and their use resulted in a tree's jumping cleanly from the stump, thus minimizing waste by breakage. (Many stumps that still exist, firm and sound after almost a century, exhibit only the marks of the fallers' axes, none of saws. To picture the long and tedious task of cutting down a Coast redwood with axes only is to understand how the word "choppers" originated, and why it is still more commonly used than "fallers.") Driver holes were now cut into the tree a few feet from the point selected for the undercut, and drivers inserted to support the staging boards on which the choppers would stand. The undercut was made on that side of the tree toward which it was to fall. Both choppers worked together, using double-bitted axes of three to four pounds in weight, with strong and elastic hickory helves from thirty-eight to forty-two inches long. Because they had to reach four or five feet into the tree to complete the undercut, these were the biggest and heaviest axes used in the Unted States.[14]

On the opposite side of the redwood, the loggers cut a ring around the trunk through the thick bark, slightly above the bottom level of the undercut, marking the point at which they would start to saw with a crosscut saw. As the saw bit steadily into the tree, numbers of steel wedges were inserted to remove pressure from the blade, as well as to steer the tree in the right direction, a process which continued until the redwood went down, driven by the force of the wedges themselves.

To fell a redwood properly, thereby avoiding breakage waste, called for consummate skill on the part of the woodsmen. In the early days of redwood logging, twenty-five percent of each tree — the leafy unused top and the tall stump to be left behind — had to be written off as waste even before the cut was made. Near the bottom of every redwood's trunk was a large swelling that was too awkward for the mill to handle; hence the choppers had to cut into the tree above that protuberance. In addition, breakage from felling, swamping, and decay ran as high as ten percent more — to the point where only about sixty-five percent of any felled tree reached the mill in the form of logs. Improved methods, better machinery and equipment, and above all scores of choppers or fallers who had become not only knowledgeable about and sensitive to the characteristics of redwood, but highly skilled in handling their implements, combined to hold breakage and waste to a minimum.[15]

Once a redwood had been felled, a sawyer would examine it and determine how it could be sized to best advantage. Logs were cut in even lengths from twelve to twenty feet. Rarely in the late 1800's would one exceed the latter figure. A ring would be cut into the trunk at the spot or spots where it was to be sawed, and peelers would begin the difficult and dangerous chore of removing the bark with flattened iron bars. The buckers, as the men who cut the lengths were called, had acquired long crosscut saws with curved bellies ranging from eight to twelve feet in length, and things had changed in that they did not, as in earlier years, have to file and set their own saws, as well as saw. By the 1890's most logging camps employed a filer to keep all the saws sharpened. Nevertheless, the sawyers maintained a hearty respect for their equipment, and a good bucker would usually continue to insist on using only the particular saw he was used to.[16]

Old-timers tell so many stories about "the big, big trees" in the "good old days" that it becomes well-nigh impossible to disentangle fact from fiction. Nor is it terribly important to do so, as long as the "fiction" reinforces the facts. However, this study is not the place for tall tales. So for the general reader, as well as the expert on redwoods, here are some fairly hard data:

Two references have been found, both from the 1850's, to a huge goose pen redwood, 33 feet in diameter, that was located between Trinidad and the Klamath River near Redwood Creek and used by packers for camping quarters.[17]

In November 1886, the Elk River Mill and Lumber Company felled a tree and cut it into 21 logs: 4 logs 16 feet long; 12 logs 20 feet long; and 5 logs 24 feet long. Total length end to end equalled 424 feet, and the logs produced 79,736 board feet of lumber.[18]

At about the same time, there was a huge tree cut in the Vance Woods on Mad River. That one was 275 feet long (and 50 feet to the first limb!). The first log off it was 17 feet in diameter at the butt and 16 feet long; the third log was 12 feet in diameter at the end. In 31 feet, the trunk of the tree diminished only 5 feet.[19]

On January 28, 1905, *The Humboldt Times* ran a feature headed "Story of the Big Tree":

> A redwood tree blew down in Section 34 T & N, R 1 E HM, a short time ago on the lands of the Dolbeer & Carson Lumber Company. It was left by the loggers because they could not fell it and save it. It broke at 130 feet, where it measured 19 feet in diameter. It is estimated that it will cut 70,000 feet of clear lumber, 60,000 feet of merchantable, 150 cords of shingle bolts and 100 cords of wood. A splinter was slivered from the tree when it fell and will make about ten cords of shingle bolts.

On July 24, 1911, the editor of *The Humboldt Standard* published an article in that paper voicing his grave concern that the "Biggest Redwood in the World" was soon going to be felled by woodsmen employed by the Northern Redwood Lumber Company. This tree, situated just east of Korbel, was estimated to be thirty feet in diameter and three hundred feet tall. The citizens of nearby Blue Lake wanted to preserve the "Big Tree of Canyon Creek" as a landmark. It grew in a protected area, and they were convinced it would stand for the "next couple of centuries." But the giant went to its inevitable fate — death at the hands of the woodsmen.

In June 1956 the Simpson Logging Company felled a tree in the Big Flat area on the north fork of Mad River that was 19 feet 6 inches inside the

In the Riverside Lumber Company woods in the 1890's, bull buck Norman Graham and a chopper lie on the stump to show the diameter and depth of the undercut. Chopper Charlie Johnson is on the left, and chopper Joe Mills is on the right. — A.W. ERICSON

Choppers take a breather in the Hammond Lumber Company woods in the early 1920's after completing the undercut. The smaller tree to the right has to be felled and removed before the back cut is completed on the larger tree to avoid breakage. — OTTO AND MARY EMILY DICK — G.J. SPEIER COLLECTION

Choppers, standing on staging boards, begin the arduous work on the undercut in the Humboldt County woods in the 1920's. — PAUL JADRO COLLECTION

A chopper and a Hammond Lumber Company official stand in the huge undercut of a giant redwood in the Hammond woods about 1920. — OTTO AND MARY EMILY DICK — G.J. SPEIER COLLECTION

After finishing the undercut, choppers rest before working on the back cut in the Mendocino Lumber Company woods in the 1870's. — ROBERT J. LEE — CARPENTER COLLECTION

Woodsmen and their children pose in the undercut of an 18-foot diameter redwood near Camp 20 in the Hammond Lumber Company woods in the early 1920's. — OTTO AND MARY EMILY DICK — G.J. SPEIER COLLECTION

Two unidentified choppers pose in the undercut of a 14-foot diameter redwood in the Humboldt County woods before working on the back cut. — CLARKE MUSEUM COLLECTION

Bull buck Norman Graham (seated on the left) with choppers Charlie Johnson (standing on the left) and Joe Mills (far right) in the Riverside Lumber Company woods in the 1890's. The other chopper, seated with Graham, is unidentified. — A.W. ERICSON COLLECTION

Woodsmen, in the Northern Redwood Lumber Company woods in 1906, pose in the undercut of a 16-foot diameter redwood. — PETER PALMQUIST — J. MASEMAN PHOTO

Three choppers, after completing the undercut in the Hammond Lumber Company woods in the 1920's, pose with the gunstick which is used to determine the direction of the fall of the tree. — OTTO AND MARY EMILY DICK — G.J. SPEIER COLLECTION

Choppers with gunstick after completing the undercut in a giant redwood in the Hammond Lumber Company woods in the early 1920's. — OTTO AND MARY EMILY DICK — G.J. SPEIER COLLECTION

Photographer A.W. Ericson's famous photo of choppers working on the back cut in the Vance woods near Fieldbrook in Humboldt County in the late 1880's.

Choppers in the Humboldt County woods pose for the photographer just before completing the back cut to send the tree toppling forward. — SAM SWANLUND

Choppers stand on staging boards, in the Little River Redwood Company woods about 1915, just before finishing the back cut to make the giant redwood fall to the left on the undercut side. — GEORGE KNAB COLLECTION

Timmberr! Arcata Redwood Company choppers, in the Arcata Redwood Company woods, watch as another redwood falls on a prepared bed (leveling of ground) to break the fall.
— DAVID SWANLUND

Mr. and Mrs. Salmon of Havana, Cuba, representatives of the Little River Redwood Company for the West Indies, pose with woods boss, John Sequist, in the company woods in the 1920's after a huge redwood was felled. — OTTO AND MARY EMILY DICK — G.J. SPEIER COLLECTION

Hammond Lumber Company loggers have their picture taken in the Big Lagoon woods in 1939. From left to right: Floyd Johnson, hooktender; Bill Boyle, rigger; and Clyde Terry, high climber. — KATIE BOYLE COLLECTION

Two choppers, after felling a redwood in the Hammond Lumber Company woods in the early 1920's, pose with their tools. — OTTO AND MARY EMILY DICK — G.J. SPEIER COLLECTION

Ringers and peelers at work in the Hammond Lumber Company woods about 1920. The peelers followed the ringers, peeling the bark off the logs before they were bucked into log lengths by the buckers. — OTTO AND MARY EMILY DICK — G.J. SPEIER COLLECTION

The ringer at work in the Hammond Lumber Company woods about 1920. The ringer used to cut rings around a felled redwood through the bark at intervals along the log with a razor-sharp ax to help the peelers in their work — OTTO AND MARY EMILY DICK — G.J. SPEIER COLLECTION

An early photo of woodsmen at work after felling a big redwood in the Humboldt County woods in the 1870's. The log has been ringed and is ready for the buckers. — A.P. FLAGLOR PHOTO — PETER PALMQUIST COLLECTION

Peelers removing the thick fibrous bark from a redwood log with peeling bars in the Little River Redwood Company woods near Crannell about 1915. — GEORGE KNAB COLLECTION

Hammond Lumber Company peelers start to peel a redwood log, working from the top and down around the log. After the logs were peeled, they were bucked into log lengths by the buckers. — OTTO AND MARY EMILY DICK — G.J. SPEIER COLLECTION

Peelers at work in the Hammond Lumber Company woods about 1920. Redwood logs were peeled from the top down to the ground. Peeling was hard and dangerous work. — OTTO AND MARY EMILY DICK — G.J. SPEIER COLLECTION

Felled and peeled logs in the Excelsior Redwood Company woods in the Freshwater area of Humboldt County in the 1880's. After the logs were bucked, the area was burned over before the logs were yarded out. — A.W. ERICSON

Loggers bucking a big redwood log for shingle bolts in the Vance woods in Humboldt County in the 1880's.
—A.W. ERICSON

A chopper and powder monkey pause for a moment before inserting dynamite to split a huge redwood log. — PAUL JADRO COLLECTION

An unidentified bucker bucking a redwood log in the Hammond Lumber Company woods about 1915. — OTTO AND MARY EMILY DICK — G.J. SPEIER COLLECTION

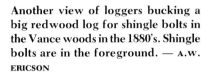

Another view of loggers bucking a big redwood log for shingle bolts in the Vance woods in the 1880's. Shingle bolts are in the foreground. — A.W. ERICSON

Huge logs cut from one tree in the Vance woods near Mad River in the late 1880's. The logs were from 13 to 19 feet in diameter, and the tree was approximately 300 feet long. Charlie Hitchings, timber cruiser, is standing in the foreground. — A.W. ERICSON

The reliable oxen pulling a train of logs over a corduroy road in the John Vance & Company woods near Mad River in Humboldt County in the 1880's. — A.W. ERICSON

An early wooden railroad in Mendocino County using horsepower. The trucks were fitted with big axles and large flanged wheels hewn to fit the wooden rails. — CALIFORNIA REDWOOD ASSOCIATION

Oxen bringing a train of logs to be dumped into Elk River in Humboldt County. Splash dams were constructed across streams to build up a head of water for moving logs in the early days of logging. The man in the foreground has a jack screw to move the huge logs. — SETH BUCK COLLECTION

Champ Clark Tree

James B. "Champ" Clarke served in the U.S. House of Representatives for 26 years, and from 1911 to 1919 was Speaker of the House. In 1915, while in San Francisco, he traveled north to see the redwoods. This picture was taken in an undercut of a 19-foot, 9-inch redwood in the Hammond Lumber Company woods. The men, from left to right, are as follows: (top row) Lawrence F. Puter, Champ Clarke, H.L. Ricks, Sr., Thomas Cutter, Will N. Speegle, James F. Coonan, George W. Fenwick; (seated left to right) Harry Hine, Joe Freitas, H.W. Hamilton, Unknown, C.P. Soule, Walter Coggeshall, Simeon Fraser, George A. Kellogg, Mike Williamson, Dr. Wattentaugh, Frank Laughlin (woods boss). — S.V. BUNNELL PHOTO — CLARKE MUSEUM COLLECTION

This 20-foot diameter section, which was used for publicity purposes by the John Vance Mill & Lumber Company and the Eureka Chamber of Commerce, was cut in the Vance woods near Mad River and delivered to the mill at Samoa in 1899. It was later shipped to the Sutro Baths in San Francisco. Noah Falk, in the frock coat at the left, was a key man in many of the early mills. — A.W. ERICSON

Hammond Lumber Company woodsmen of Camp 20 (LEFT) have their picture taken in front of a 20-foot diameter redwood on a Sunday in 1925. — HENRY SORENSEN COLLECTION (CENTER) The town of Scotia claimed the largest stump in Humboldt County. This stump, which was 32 feet in diameter, was used as a bandstand. — FRED ELLIOTT COLLECTION (RIGHT) This 24-foot diameter redwood, in the Big Lagoon timber above Maple Creek, belonged to The Little River Redwood Company in 1929. Captain Elam, engineer and forester, stands in the center of this view. Howard A. Libbey, assistant manager, is shown at the right buried in ferns. — GEORGE KNAB COLLECTION

A wind storm blew this huge redwood down in the Little River Redwood Company woods in 1930. John Sequist, woods boss, is on the tree, and Mr. and Mrs. Harold Salmon of Havana, Cuba are in the foreground. — GEORGE KNAB COLLECTION

A planting crew in the Union Lumber Company woods in Mendocino County in 1924. The first reforestation program in the redwoods in the 1920's failed. — CALIFORNIA REDWOOD ASSOCIATION COLLECTION

bark and close to 300 feet tall. Choppers Cooper and Devore took six hours to fell it with chain saws.[20] There was a time when choppers would have worked and sweated for days before that tree or many another like it crashed to the ground. Redwood logging has indubitably advanced to a state of total efficiency.

In 1882 the following warning appeared in a Redwood Country history book.

> The calamity which will befall the people of Humboldt County by the exhaustion of the forests of redwoods could be in a great measure averted if the growth of the young redwoods were fostered. But no care is taken; and, in fact, it seems that an effort is made to thoroughly eradicate all traces of the forests. The stumps are fired just to see them burn, and fire runs over the land every fall, which serves to completely destroy the young shoots. The protection of our forests should be a charge of our Legislature; for, while the men of to-day may not remain to suffer for the want of these forest trees, the commonwealth of the State will remain, and its future wealth should be cared for by the present generation.[21]

The great age of the redwoods made many timber owners feel that growing a new crop of them would be an impossibility. Much of the cutover land was converted to grass by seeding. Faced with mounting criticism of their forest management practices by the general public, however, some of the more far-sighted owners began seriously considering the possibilities for reproduction. Proof that it could be done was evident in the cutover lands of the bull team days, where a superb stand of second growth showed real promise for the future. One of the leaders of this movement was C. R. Johnson, founder of the Union Lumber Company in Fort Bragg. In the 1920's, other lumber companies joined him in a massive "reforestation by planting" program. The project continued for about eight years, but was then abandoned. The companies spent a quarter of a million dollars, only to find that redwood seedlings were not as hardy as those of other species, in spite of the tremendous vitality and high resistance to unfavorable conditions of mature redwoods.[22]

Not until 1950 did the first tree farm, an area of forest land in private ownership devoted primarily to the continuous production of commercial timber crops under good forest practices, come into being. Designed to assure a future supply of commercial timber crops under good forest management, it was organized by the Hammond Lumber Company under the direction of the California Redwood Association. The farm was located at Hammond's Van Duzen River unit and dedicated on August 26, 1950. Shortly thereafter the same company set up a second tree farm at its Eel River site.[23] In May 1951, the Union Lumber Company, following Hammond's example, organized two even larger tree farms at its Noyo and Big River units.[24]

An early whipsaw mill before the introduction of water and steam power. Two men had the difficult job of sawing small logs into lumber. — EUREKA CHAMBER OF COMMERCE

Woodsmen relaxing on the porch of their shanty in the Mendocino logging camp in the 1890's. — CALIFORNIA REDWOOD ASSOCIATION

Chapter Notes

1. *The Humboldt Times*, December 31, 1899.
2. Peter Rutledge, private interview, Eureka, California, March 22, 1955.
3. John Larson, *Oral History Interview with P. J. Rutledge, Eureka, California, March 1953* (St. Paul, Minn.: Forest History Foundation, Inc.).
4. David R. Leeper, *The Argonauts of Forty-Nine* (South Bend, Ind.: J. B. Stoll & Company, 1894), 135-136.
5. *The Humboldt Times*, September 9, 1854; Lindsey, "Statement of Reminiscences," in Owen C. Coy, *The Humboldt Bay Region 1850-1875* (Los Angeles: The California Historical Society, 1929), 287.
6. Lillie E. Hamm, publisher, *History and Business Directory of Humboldt County* (Eureka; *Humboldt Daily Standard*, 1890), 51.
7. Peter Rutledge, private interview, Eureka, California, June 5, 1956.
8. *The Humboldt Times*, 85th Anniversary Edition, E-2.
9. David Warren Ryder, *Memories of the Mendocino Coast* (San Francisco: Taylor & Taylor, 1948), 16.
10. Emanuel Fritz, private interview, Eureka, California, March 22, 1963.
11. Hamm, *History and Business Directory*, 55.
12. Harry Ryan, private interview, Arcata, California, March 15, 1960.
13. Fritz, private interview, March 22, 1963.
14. Hamm, *History and Business Directory*, 52: Rutledge, private interview, June 5, 1956.
15. California State Board of Forestry, *First Biennial Report, 1885-86*, 139.
16. Rutledge, private interview, June 5, 1956.
17. Leeper, *Argonauts*, 144.
18. *The Humboldt Times*, November 18, 1886.
19. Howard Brett Melendy, "One Hundred Years of the Redwood Lumber Industry, 1850-1950" (Ph.D. dissertation, Stanford University, 1952), 35.
20. *The Humboldt Times*, June 10, 1956.
21. Wallace W. Elliott, *History of Humboldt County, California* (San Francisco: W. W. Elliott & Co., 1881), 141.
22. Emanuel Fritz, "The Changes I Have Seen," *The Humboldt Times*, The Redwood Logging Conference Silver Anniversary, March 7, 1963; Melendy, "One Hundred Years," 262-264.
23. *Welcome to Hammond Lumber Company*, pamphlet prepared by the former Hammond Lumber Company.
24. Melendy, "One Hundred Years," 272.

Fallers using an early chain saw in the late 1940's to fell a 15-foot diameter redwood in the Arcata Redwood Company woods in Humboldt County. — DAVID SWANLUND

A Tale of Two Saws

Redwood Country timber was still being felled by hand as late as 1935, but efforts to replace manpower in the woods with machine-driven saws began much earlier. In October 1875, John Dolbeer, an inventive genius who later patented the Dolbeer Steam Logging Donkey and the gypsy locomotive, invented a portable power saw for use in felling trees and bucking logs. It included an upright boiler, mounted on wheels for easy moving, and an adjustable engine that rested on a stout frame. A rubber hose carried steam from boiler to engine, so the boiler could remain in one spot while the engine was moved to any nearby log[1] The saw itself was fastened to a piston rod and fed by moving the cylinder with a screw. Before cutting a tree, fallers would turn the engine on its side and secure it to a log by means of iron rods. Before cutting logs, buckers would place the engine upright so that the saw would cut downward. Although its trial runs were successful, no evidence exists to show that this Dolbeer saw became part of the lumber industry's standard equipment.

For a time, most new or improved power-driven saws were variations of the drag saw or pitman type of machine, run by gasoline, steam, or air. The first drag saw (a crosscut saw powered by a small gas engine) in the Redwood Country was built at the Eureka Foundry in August 1886 for Swortzell & Williams of Springville (Fortuna). It was believed capable of accomplishing the work of three men in turning out shingle bolts.[2] Of the many subsequent varieties of drag saw, four — the Wade, the Vaughn, the Hansen, and the Little Woodsman — covered the spectrum of weights and sizes produced by local manufacturers.[3]

Before 1920 an unlimited market for split products (posts, grapestakes, shingle bolts, etc., split from trees where they fell) existed, and in those days, the Wade and other drag saws of the 300-pound class were used primarily in making them. Efforts on the part of lumber companies to use the Wade for felling were thwarted by the old-timers, who fought such changes and insisted that with its heavy weight and inadequate engine, the saw wasted more time than it saved. The Wade was used for bucking, however, and later improvements to it aimed at increasing its efficiency for that operation.[4]

Lumber companies continued to be enthusiastic about powered saws and confident they could increase production and reduce costs by using them. In 1935 The Pacific Lumber Company, which was using drag saws for both felling and bucking in the Freshwater area, reported a forty percent

Two unidentified scenes of loggers experimenting with an early day drag saw near Eureka's Sequoia Park about 1903. — JACK BUEHLER COLLECTION

ESTABLISHED 1910

HANSEN MACHINE & WELDING WORKS
COMPLETE REPAIR SERVICE
MANUFACTURERS OF LIGHTWEIGHT DRAG SAWS

PHONE EUREKA 78
7TH & BROADWAY
EUREKA, CALIF, U.S.A.

Falling Timber with a Eureka Air-Cooled Drag Saw

SHORT FRAME

First take out wingbolt on left frame handle removing handle. Loosen right frame handle clamping bolt, jerk forward to end of slot in handle, and fold handle around over treebolt and bracket or remove.

Next drive dragsaw hanging wedge in tree from 3 to 6 inches from gun stick mark, or desired depth of under cut (either side of tree). Hang dragsaw to wedge by slot in left iron box on end of frame. Drive tail dog in tree opposite the undercut if possible, below level of stump, attach to eye bolt in frame, line sawbolt with layout and fasten tail dog with set screw. Then swing lower side of frame out to saw a Vee block for undercut, level frame with tree bolt provided. Drive in dog attached to frame, swing hold up rod around under guide rods, and fasten on the blade.

Drive rope dog in tree above and behind dragsaw, fasten lever.

Directions from the Hansen Machine Company, early Eureka pioneers in the drag saw business, on how to operate an air-cooled drag saw. — THE PACIFIC LUMBER COMPANY COLLECTION

reduction in operating costs. The same year, Hammond Lumber Company introduced the Wade saws in their woods and used them for felling and bucking. The better quality work that resulted was soon evident to management, and labor welcomed the consequent increase in income. In 1936 Hammond doubled its mill production goals and, in order to meet them, increased the number of fallers working in its Big Lagoon area from sixty-five to ninety-three. The entire lumber industry was undergoing a general expansion about this time, and as a result, many loggers encountered their first experience with a power felling and bucking.[5]

By 1937 drag saw felling and bucking had become the established practice, and by 1939 over eighty-five percent of all felling and bucking was done by this method. Only on steep slopes were these operations still carried out by hand. The heavy Wade saw was soon replaced by the lighter Vaughn saw. The Vaughn weighed 225 pounds, and designers continued trying to reduce its weight to the point where it could be used on steep slopes. Indeed, all the major lumber companies experimented with various drag saws, hoping to increase efficiency.[6]

By 1941 Harold Hansen of Hansen Machine Works in Eureka had developed the Hansen saw, a lightweight air-cooled model with special felling and bucking devices. Five years later, he came up with the Hansen Pony. This weighed approximately 150 pounds, or two-thirds as much as the original Hansen. By 1947 there were 2,100 Hansen drag saws operating in the woods.[7]

Early in the 1940's, Cliff Merrill decided to build a lightweight saw that he himself, a small man, could handle easily. In preparation he went into the woods to learn about felling and bucking, and then worked for two years in a machine shop in order to master the machinist's trade. In 1944,

A faller uses an early drag saw in the 1920's to fell a redwood. — PAUL JADRO COLLECTION

Fallers experimenting with an early drag saw in The Pacific Lumber Company woods. — THE PACIFIC LUMBER COMPANY COLLECTION

Bucking a redwood log with a gasoline drag saw in the Freshwater operations of The Pacific Lumber Company. — THE PACIFIC LUMBER COMPANY COLLECTION

Veteran chopper Dick Beach demonstrates the drag saw which he used in the Hammond Lumber Company Big Lagoon woods in the 1930's. — KATIE BOYLE COLLECTION

working with a Mr. Barnwell, of Eureka, Merrill put on the market a 95-pound air-cooled drag saw called the Little Woodsman. The Little Woodsman came equipped with a light saw blade, designed to reduce weight and manufactured by the Simmon Saw Company; but its motor was powerful enough to pull a bigger conventional blade, and the Little Woodsman Company furnished an adapter for that purpose. Thus the saw could operate with the 20-foot blade being used by most of the lumber companies at the time, and by 1949 Merrill had made and sold 2,500 Little Woodsmen.[8]

During the period when the initially heavy and cumbersome drag saw was developing into an efficient lightweight machine, chain saws began to appear in the woods. The first experiment with a chain saw took place in 1905 in Henry Medle's greenhouse near Sequoia Park in Eureka. The name of its inventor is unknown. Driven by gas, with a two-cylinder water-cooled system suspended on a tree above the engine, the chain saw's teeth were set singly on a flexible chain, similar to an over-sized bicycle chain, that ran around a long oval bar. Later manufacturers tried making chain saws powered by electricity or by compressed air.[9]

The first chain saws used in the Redwood Country were known as scratch or barbed wire chains. Showers of fine particles thrown up by their teeth delayed operations. They cut slowly, and filing them was a tedious task. They also vibrated terrifically, demanding extra effort on the part of the operator, who therefore tired more rapidly. These chain saws were poorly received by most fallers, and their engines broke down often because the men did not know how to operate or maintain them properly.[10]

On the end of the bar on early chain saw models were a pulley and stinger that limited the diameter of trees they could cut to just under seven feet (the length of the bar). This limitation was a serious drawback, because to cut a redwood ten or twelve feet in diameter, fallers had first to side notch it so that neither the undercut nor the backcut would exceed seven feet.[11] Side notching was worse than just a time-consuming procedure. When a tree with a side lean is notched, its pulling power —that characteristic of the tree which "steers" it in the direction loggers wish it to fall — is lost, since that power resides in the tree's sapwood or outer portion. Many such redwoods were broken in the course of being felled by the first chain saws, and quantities of valuable wood were lost.[12]

Buckers using the early chain saws faced a similar problem with giant redwood logs. A seven-foot bar with stinger could buck only cuts of six feet or less. If a log was bigger than six feet in diameter, it had to be side notched. In this case, side notching made some of the lumber too short, to say nothing of how difficult was the job of splitting out a slab as long as two feet to make room for the diameter of the chain saw motor.[13]

There were other disadvantages to chain saws, such as the difficulty of keeping a pinch cut open enough to allow sufficient clearance for operation of the bar and chain. A cut could be reamed only by working the saw bar up and down, which took a good deal of brute strength and resulted in many a kick back. Open cuts were worse. They had to be held open with hanging wedges, and if the hangers failed to hold and the log slipped, the saw would be firmly stuck. Nor did buckers find it easy to dislodge a chain saw after they sawed a log. Frequently they would not have enough clearance to remove the bar and stinger, and would be forced to separate the bar from the saw's transmission and draw it out from the side opposite the machine, an operation that often called for wedges. Before they could make the next cut, they would have to reassemble bar and chain and reconnect them to the transmission and drive sprocket.[14]

So much breakage and waste resulted from improper operation of chain saws, as well as from

Bruno Falleri working on the back cut in the Simpson Timber Company woods demonstrates a new chain saw. — DAVID SWANLUND

Modern-day choppers finishing the undercut with a chain saw, much different from the huge undercuts once chopped by hand in the early days. — DAVID SWANLUND

their inherent failings, that lumber companies began restricting their use to straight felling and small timber that contained a high percentage of fir. A chain saw used to cut small timber became known as "the scaler's nightmare," since trees in a piled-up condition were next to impossible to measure accurately, and logs usually had to be scaled twice. Also, the racket made by chain saws was dangerous to a scaler. It prevented him from alerting fallers to his presence, a hazard that increased greatly when a scaler was in steep country, working below a set of fallers.[15]

Early in 1949, however, the chain saw was tremendously improved by the invention of the bucking bar, a saw with no stinger and a round-tipped bar. A saw equipped with a five-foot bucking bar could fell a redwood as big as ten feet in diameter, since it could be driven into a tree head-on. It could also bore an undercut in such a way that the wedge could easily be split out, and it could quickly bore through or rip down suckers, to facilitate felling the big trees.[16] These new and better chain saws began taking over from drag saws, in spite of the reduction of weight that had been achieved in the latter.

A present-day chopper still uses the gunstick to sight the fall of a redwood in the Simpson Timber Company woods near the Klamath River. — DAVID SWANLUND

A modern-day faller, on staging boards, using a gasoline-powered chain saw in the late 1960's in the Georgia-Pacific Big Lagoon operations. — DAVID SWANLUND

Nevertheless fallers still had a lot to learn before they could operate the bucking bar at maximum efficiency. Improper steering surfaces caused the loss of many redwoods. Corners would be sawed too close or sawed off, or too much wood would be left in the center of a tree. Manufacturers tried to eliminate such problems by making longer saw bars, and for a number of years, saws could be fitted with bars of eight, ten, and even twelve feet. The longer the bar, however, the harder the chain saw was to handle, especially in steep slope felling. Longer bars also required extra horsepower because of their wide cutting surfaces, and friction built up between bar and chain. Filers had to be expert at their task to prevent bars from becoming plugged up with sawdust. Overall, the longer bars proved slow-cutting, cumbersome, and hard on a chain saw's motor and clutch.[17]

At first, because of the poor work that had been turned out with the short bar, lumber companies extended the life of the long ones. The better redwood fallers and buckers, however, began proving they could do a good job using the short bar, with less effort and in less time. Improved chains also became available, and competition among sawmakers resulted in the design and production of lighter-weight machines. Within two years of the invention of the bucking bar, a chain saw in the hands of a competent logger had become the most practical tool for both felling and bucking the redwoods.[18]

A Hammond Lumber Company faller using a chain saw to fell a redwood in the Big Lagoon operations in the 1950's. — DAVID SWANLUND

By the 1950's, many stands of big trees in the Redwood Country had been cut, and lumber companies were logging smaller timber on steep slopes farther and farther away from stream beds. That kind of logging did not lend itself well to drag saw felling. In bucking, if a log was to be undercut with a drag saw, the logger first had to dig a hole under the log and frequently open it up with powder before a short drag saw blade could start cutting. Longer drag saw blades saved little time or effort, since they required changing as cutting progressed. Drag saw bucking called for propping butts and tops and holding cuts open by means of hanging wedges. When a drag saw was doing the bucking, roughly ninety percent of the cuts had to be finished off by hand. Before a drag saw could fell a large tree, the tree's suckers had to be removed, frequently with blasting powder or dynamite.[19]

The chain saw, on the other hand, was much more versatile. Undercutting a log by chain saw was a simple matter. The operator bored through the log and sawed down. Logs did not have to be undercut as deeply as they did with a drag saw,

because the chain saw could corner a top much more effectively. Use of the bucking bar eliminated the need for wedges, and almost all cuts could be completely sawed off. As for a tree's suckers, the chain saw made it possible to bypass them in many instances; but when they had to be removed, it could rip them down quickly.[20]

Perhaps the most important reason why drag saws ultimately disappeared from the woods, however, was the increased production that accompanied chain saw operations. The estimated felling and bucking output of one man in one day using a drag saw varied from 8,000 to 15,000 board feet on the Humboldt Scale. In 1951 the estimated felling and bucking output of one man in one day using a chain saw varied from 15,000 to 18,000 board feet on the Humboldt Scale. In small timber, chain saw production ran even higher than that.[21]

In spite of its bad features, such as personal hazard to its operators and the increased fire danger that accompanied its use, the chain saw eventually proved superior to the drag saw in all phases of logging operations.[22]

Chapter Notes

1. *The Humboldt Times*, February 9, 1923.
2. Ibid, August 26, 1886.
3. Bill Baker, "Drag Saws in the Redwoods," speech before the Redwood Region Logging Conference, Eureka, California, March 7, 1963.
4. Ibid.; Walter F. McCulloch, *Woods Words — A Comprehensive Dictionary of Logging Terms* (The Oregon Historical Society and the Champoeg Press, 1958), 177.
5. Baker, "Drag Saws in the Redwoods."
6. Ibid.
7. Ibid.
8. Ibid.
9. "Power Saws Come of Age," *Timberman* (October 1949): 150-151; McCulloch, *Woods Words*, 31.
10. Larry McCollum, "Chain Saws in the Redwoods," speech before the Redwood Region Logging Conference, Eureka, California, March 7, 1963.
11. Ibid.
12. Ibid.
13. Ibid.
14. Ibid.
15. Richard Beach, private interview, Trinidad, California, January 10, 1968.
16. Ibid.
17. McCollum, "Chain Saws in the Redwoods."
18. Richard Beach, private interview, January 10, 1968.
19. McCollum, "Chain Saws in the Redwoods."
20. Ibid.
21. Ibid.
22. Beach, private interview, January 10, 1968.

Logging along Ryan Slough (Creek) in the McKay & Company woods just southeast of Eureka in the 1880's. The company built a logging railroad six miles, along the creek, into the timber. The logs were rafted from the log dump along Eureka Slough to the bay and to the Occidental mill on the waterfront between A and B streets. — ANDREW GENZOLI COLLECTION

Big Trees on the Move

In the American lumbering lexicon, "swamping" is a general term that covers the work involved preparatory to moving timber from its natural habitat to lumber manufacturing facilities. If felling the giant coastal redwoods was a herculean task, swamping them was no less an endeavor. When all the trees in the area of a particular operation had been cut and bucked, swampers built skidroads and log landings, bridges over streams, and in some cases chutes, so the logs could be moved out of the forest.

Built entirely by back-breaking manual labor, at an estimated cost of $5,000 per mile, the vital skidroad represented the most demanding and expensive aspect of early logging in the Redwood Country. Once they had shoveled and stomped and spread a roadbed into being, swampers laid skids — logs approximately fourteen inches wide and twelve feet long — at right angles to its center, forced "punching" — dirt, brush, or slabs of wood — between the skids, and then reinforced the resulting path with cross logs, to form a corduroy road on which oxen or horses could pull crude sixteen-foot trucks carrying small-diameter logs over level ground or even slight grades. Alternatively, the roadbed might serve as the base of a strap railroad, constructed by nailing scrap iron to two-by-fours placed end-to-end to make a kind of tramway, on which horses could pull wagonloads of small-diameter logs.[1]

Getting the more gigantic logs onto a skidroad, so that animal teams could haul them out of the woods, was a challenge. Some had to be lifted from their beds by means of block and tackle. At this point, the jack screw, a vital tool in the woods, came into play. The ability of two men, each equipped with a jack screw, to manipulate huge logs was little short of amazing. They could maneuver a log out of a hole and turn it completely around. Moreover, where the terrain was extremely steep, they could jack screw a log all the way down to the skidroad.[2]

Once on the road, big logs were positioned one behind another, largest in front, and fastened together with dogs (short metal stakes sharp on one end and with an eye on the other) to form a train of from three to seven logs, depending on their size.[3] Teams of eight to sixteen oxen or six to ten horses — the number determined by the weight of the logs they had to pull — hauled the train along the skidroad. Although horses were faster and could make more trips in a day, they frequently broke their legs jumping around in their traces, and they also had to be controlled by reins. Oxen

Making a train of logs (LEFT) in the Eel River Valley Lumber Company woods near Fortuna in 1903. Lloyd Scriner is in the foreground, Jerry Doughie and John Cane are on top of the logs, while Fred Happ and Elmer Richly are standing alongside the train. — FRED ELLIOTT COLLECTION (ABOVE) Another train of logs on the skidroad in the Eel River Valley Lumber Company woods in the early 1890's. — J.W. PHEGLEY COLLECTION

Preparing a train of logs in the Vance woods near Mad River in the 1880's. The lead log is at the left. — A.W. ERICSON

52

In the early days the powder monkeys had to split the large redwood logs so that they could be taken out of the woods and handled by the early mills. — SAM SWANLUND

A powder crew poses for the photographer in the Humboldt County woods in the 1880's. — CLARKE MUSEUM COLLECTION

The loggers, shown above, have their picture taken near Camp 11 of the Northern Redwood Lumber Company woods in June, 1911. P.R. Patten of Eureka is third from right (standing on log). — P.R. PATTEN COLLECTION (UPPER RIGHT) An early logging scene near Eureka in the 1860's. The man on the log is boring a hole to insert black powder to split the log so that it can be taken out in the crude ox-drawn wagon. — FREESE & FETROW PHOTO — CLARKE MUSEUM COLLECTION

Woodsmen wait on the skidroad after dogging a train of logs for the bull donkey in the Humboldt County woods. — CLARKE MUSEUM COLLECTION

A trestle built in the early days of logging near Little River in Mendocino County. An ox team pulls logs over a wooden railroad. Note the live trees forming a part of the trestle. — NATIONAL MARITIME MUSEUM — SAN FRANCISCO

The oxen on the skidroad with a train of logs in the Vance woods in the Lindsay Creek area in 1890. A logging camp and the "hump" on the road to Fieldbrook is shown in the background. — GILFILLAN PHOTO — SARAH MC CURDY COLLECTION

Loggers laying a bed for the important skidroad in the Albion woods in Mendocino County about 1890. Note the steam donkey engine in operation in the background. — CARPENTER PHOTO — ROBERT J. LEE COLLECTION

(LEFT) Loggers working on the skidroad in the Albion woods in Mendocino County about 1890. — ROBERT J. LEE COLLECTION (ABOVE) Building a skidroad in the Hammond Big Lagoon woods in the early 1920's. — LOUISIANA-PACIFIC CORPORATION

An ox team with a train of logs in the Humboldt County woods in the 1880's. — CALIFORNIA REDWOOD ASSOCIATION

Loggers pause for the photographer while working on a skidroad in the Elk River Mill & Lumber Company woods in 1907. — CLARKE MUSEUM COLLECTION

required no reins and, although they moved more slowly, were surer of foot and could also outpull a team of horses. Occasionally, when oxen were in short supply or when the skidding was unusually difficult, combinations of oxen and horses pulled together.[4]

In the more mountainous areas, logs were often chuted out of the forest. Swampers excavated a main trough up a hillside, together with a series of branch troughs that fed into it. Rigging crews rolled and pulled the logs to the nearest branch chute and propelled them down it into the main chute. As the loggers worked their way higher, swampers extended the main chute and dug additional branch chutes to keep up with them, until all the timber in the region had been sent on its way to stream bed or log landing. A key member of the rigging crew was the chute tender. In addition to patrolling the chutes, on the lookout for breaks or fires, he also had to run alongside them when chuting was under way, lessening friction and easing the logs' passage by wetting down the chute with water from tubs scattered along the route.[5]

When teams hauled logs over a skidroad, the chute tender's counterpart was the water slinger, or sugler. His duty was to accompany the team and keep the road slick by throwing water on it. The water he needed also came from tanks stationed along the road. Especially where the road was steep, a careless sugler ran the risk of sending the load hurtling down on top of the animals hauling it, or of causing the logs to stall at an awkward angle.[6] When a team started for the landing with a train of logs, the work of the water slinger began in earnest. No matter how fast the team moved, he had to keep up with it and furnish just enough water to keep the logs sliding smoothly and the animals under the least possible strain. If water was scarce, and the train got hung up, or stuck, he could expect to be roundly cursed, because a crew with jack screws would have to move in and start the logs rolling again. On the other hand, if the train moved too swiftly, the slinger had to throw dirt in front of the legs to slow them down, knowing that if he lost his footing and fell, they would run over him. The pace eased only on level stretches, where the grooved logs in the skidroad were covered with lubricating grease or tallow that smoothed the train's passage. Little wonder that the water slinger was a high-paid employee and sometimes earned more than the camp foreman.[7]

The water tubs were replenished by a water packer, or water buck, and in many logging camps, this job was performed by a boy. In a personal interview, old-timer Peter Rutledge de-

The ox team, shown above, with a train of logs on a skidroad cut through an abandoned railroad in the 1880's in Humboldt County. — A.W. GILFILLAN PHOTO — CLARKE MUSEUM (LEFT) A chute in the Hammond Lumber Company woods in the early 1920's. The log at the top is about to be chuted down the hill. — OTTO AND MARY EMILY DICK — G.J. SPEIER COLLECTION

A horse team in Isaac Minor's woods near Glendale in
Humboldt County. The two lead horses were killed by
a rolling log minutes after this picture was taken. —
A.W. ERICSON

A horse team bringing in a train of logs to the skidroad
at Greenwood Creek in the Mendocino County woods.
— ROBERT J. LEE COLLECTION

Newell's horse team bringing in a huge log into the landing at the Eel River Valley Lumber Company woods in 1893. Loggers in the background pause for liquid refreshment while the photographer adjusts the focus on the camera. — FREESE & FETROW PHOTO — WALLACE MARTIN COLLECTION

scribed his experiences:

> I was fifteen years of age in 1889 when I went into the woods. My job was what was called a water buck. Water was used on the chutes and on the skidroads. Tubs were staggered on both sides of the skidroad or the chutes. The water buck had a horse, sometimes a mule, with a pack saddle on it. And on each side of that pack saddle a canvas sack was hung, which would hold about fifteen gallons of water.
>
> We would lead the pack animal to a spring or creek to get water, and we would fill one of the sacks on one side halfway. The old horse was trained to turn about, and we would completely fill the sack on the other side. The animal would turn again and we would finish. Then we would start up the hill. The bottom of the sack was heavy sole leather and there was a wooden peg in the bottom of the sack. The trained animal would stop at a tub, and we would pull the peg out and allow a quantity of water to flow into the tub. As we proceeded up the road or chute we distributed the water into the tubs.
>
> I had a long chute, about three-quarters of a mile. And you can imagine how only a boy, and a pretty husky boy at that, could stand that kind of work. It kept me busy all day to keep the supply of water — up and down these hills all day — and we worked about eleven hours a day. We had a twelve-hour day, but part of the twelve hours was taken in going out from the camp to the operations and back again. But, anyhow, us water bucks went up and down those hills all day long — up with a load of water and down for another load![8]

Danger accompanied the bull puncher, or bull whacker, who yelled, screamed, and swore as he drove his team, reinforcing his colorful language with a sharp stick, called a goad stick that he used on the animals. Maneuvering a train of logs down a very steep hill, he was apt to find himself on the verge of being pinned against a bank. When this happened, he would leap atop the oxen at the last possible second, run across their backs, and continue leading them down from the other side — still yelling and cursing! Bull whackers usually wore tennis shoes, for quicker and surer footing. They controlled the animals by commands: "Get up," to go ahead; "Whoa," to stop; "Gee," to turn right; and "Haw," to turn left. The strong lungs of the bull puncher were no small factor in the development of the lumber industry, because some of the logs that had to be hauled were enormous. One train of seven logs hauled in the Humboldt area in 1887 by teamster A. A. Marks, with five yoke of oxen, scaled 22,500 board feet of merchantable lumber.[9]

The first logging railroads appeared in the Redwood Country in the 1870's. Logs no longer had to be hauled all the way to a stream bed for floating to the mill. Instead, they were yarded at the landing, and there loaded onto railroad logging trucks. At first loggers using jack screws loaded these trucks. After 1881, however, the loading operation was carried out by means of a steam donkey and block and tackle.

From the log landing, trains of twelve to fourteen cars, usually pulled by a Baldwin locomotive, headed for the log dump. At the dump, the blocks and chains that secured the logs to the cars were removed, a jack screw was put to work, and in a matter of minutes, the entire trainload lay in the slough, ready to be made into rafts. A stern-wheel steamer towed each raft down Humboldt Bay to the mill, where the raft was broken up and a boom guided the logs into the mill's log pond.[10]

Oxen and horses provided the main source of power for moving logs in the woods until the 1880's; then, in 1881, the invention of the donkey engine revolutionized logging operations in the United States. On July 31, 1881, the *Eureka Humboldt Times* reported:

> To those who are familiar with the logging woods, or who have been accustomed all their lives to see the unwieldy ox-team tugging at the great logs, the announcement that steam power is to supersede this antiquated method will be received with no little surprise and incredulity.
>
> But improvement is the order of the day. There is no reason why the ox-team should not make way for the steam engine, as the stage coach has for the engine.
>
> George D. Gray of the Milford Land and Lumber Company has brought up from San Francisco and set to work on the company's logging claims at Salmon Creek a steam engine that bids fair to change very considerably our system of logging. The idea of such a contrivance was laughed at by practical men, and when the engine was set down among the crew at Salmon Creek, it was pronounced utterly impracticable.
>
> But the inventors had faith in the new machine and put it in motion last week. It has performed its work to perfection from the first, and is now a regular hand in the Salmon Creek woods. The machine is designed to be used in blocking out roads, hauling out of the way all waste material and hauling logs into the roads, and coupling them together ready for the ox-team to take away. It consists of an upright boiler and engine. The crank shaft of the engine is geared into a "gypsy head." The whole is set on a heavy wooden frame, twelve feet long by six feet wide.
>
> The gypsy projects over one side of the frame. The weight of the whole machine with water and fuel is about four tons, and it has power to

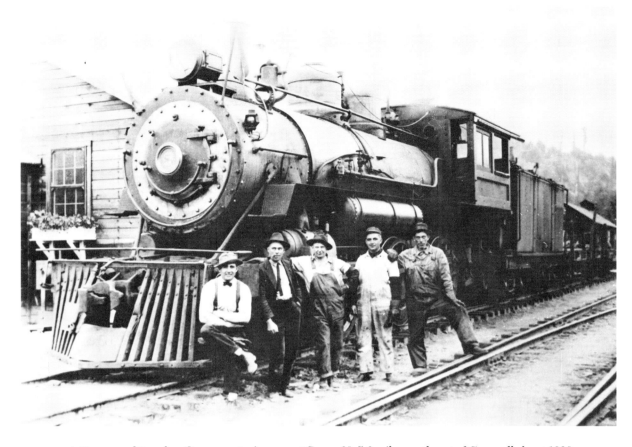

A Hammond Lumber Company train crew at Camp 20, 5.2 miles northeast of Crannell about 1926. The train crew (left to right) are Jack McGowan, brakeman; Tom Jordan, conductor; Brig Fields, brakeman; unidentified brakeman; and Henry Ohlendick, veteran engineer. — JACK TREGO COLLECTION

sufficiently break a four and a half-inch manila rope.

The operation of the machine is very simple. After running the rigging the same as for an ox-team, a few turns of the line are thrown over the gypsy and the engine is started. Besides being more powerful than the ox-team, the power can be used or halted instantly.

A log can be rolled up on its side and held there by the break to be peeled or "snipped" or for any other work that may be necessary.

The machine is the invention of Messrs. Dolbeer and Carson.

John Dolbeer, of the Dolbeer & Carson Lumber Company, patented his new invention on April 18, 1882. According to Walter McCulloch's dictionary of logging terms, *Woods Words*, Dolbeer "adapted a ship's capstan for his logging rig, and it is possible that he also brought along the usual seafaring term for the engine itself."[11] Dolbeer's first engine consisted of an upright boiler with a single cylinder that turned a gypsy head on a horizontal shaft. In 1883 he received a patent for an "Improved Logging Machine." On this new donkey, an upright spool replaced the gypsy head. Both types of engine were used throughout the Redwood Country until about 1910.[12]

A steam donkey crew consisted of an engineer who operated the steam engine, a spool tender who handled the manila line, and a boy who supplied the boiler with wood and water. The spool tender gave the 4-½-inch manila line two or three turns around the upright spool, and then played it out to the log. Block and tackle augmented the line's power to move logs into position. As the donkey engine pulled the long line, the spool tender played the excess onto the ground, instead of allowing it to wrap around the spool. Line horses, waiting nearby, then pulled the line back out to the woods. The donkey crew was under the supervision of a head chain tender, whose main responsibility was to make up the trains of logs to be moved out from the landing by teams.[13]

The common practice of employing two donkeys to a skidroad caused a lively competition between their crews, who strove to surpass each other in the quantity of logs they got out. The record was supposedly achieved by Howard Jorden's crew in 1914. Working for the Dolbeer & Carson Lumber Company in a gulch near Lindsay Creek, a short distance above Fieldbrook, the crew got out 22,400,000 board feet of lumber in ninety working days — an average of 249,000 board feet per day![14]

An early logging scene near Greenwood in Mendocino County. The Dolbeer donkey has pulled in the logs from the hillsides to the skidroad where the loggers have prepared a train of logs for the ox team. — NATIONAL MARITIME MUSEUM — SAN FRANCISCO

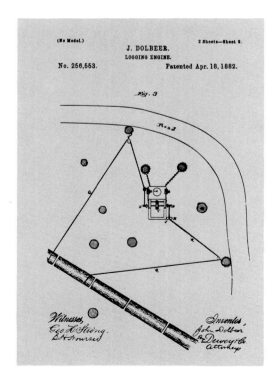

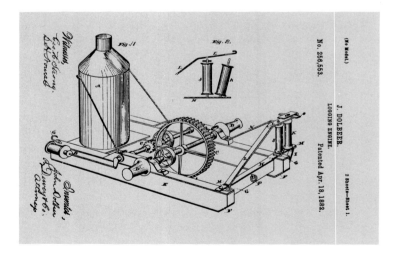

Patent drawings of John Dolbeer's logging engine awarded by the U.S. Patent Office on April 18, 1882. The apparatus was designed to move heavy logs after trees had been felled and cut up. (ABOVE) A view of the apparatus with an insert of the guide rollers separated to admit a rope cable. (LEFT) A general plan of the apparatus and operative ropes.

The early Dolbeer donkey engine at work in the Albion woods in Mendocino County. Bill McCormick is the logger in the left-hand corner. — NANNIE M. ESCOLA COLLECTION

An early Dolbeer donkey engine, shown below, brings in a redwood log in the Humboldt County woods. — JACK BUEHLER COLLECTION

Photographer A.W. Ericson's popular photo of loggers using the Dolbeer donkey with the vertical capstan and manila rope to bring in logs in the Vance woods near Mad River in the late 1880's. The water packer is in the foreground. — MRS. PERCY J. BRYAN COLLECTION

An early logging scene in the Humboldt County woods in the 1880's. A Dolbeer donkey with a vertical capstan is bringing in logs with manila rope. — CLARKE MUSEUM COLLECTION

A noon break in The Pacific Lumber Company woods about 1895. — DAN WALSH COLLECTION

As the 1880's progressed, logging railroads began moving logs faster than animal teams could, and woods crews had to provide more logs in less time, in order to keep the trains loaded. The Dolbeer Steam Donkey, by pulling in heavy logs too big to be hauled by oxen or horses from ravines and hillsides, enabled them to do so.[15] Nevertheless, the donkey at first proved a less than ideal innovation. Many a hemp rope broke before loggers realized that steel rope could do the job better, and that engine and winch should be specially designed to suit a particular logging operation. Once the machine's potential became evident, however, improvements and refinements followed rapidly, and each donkey manufactured surpassed its forerunner in capability. Steam power, introduced by pioneer Humboldt loggers, became the mainstay of woods operations for many years, and also opened the door to numerous other operations involving wire cable, among them high lead logging.

In 1883 John Dolbeer installed an unmilled cog wheel drive on a logging locomotive and came up with the gypsy locomotive, a new and efficient machine for loading logs at the landing, on which a central cog ran both the wheel drive (geared to neutral, forward, and reverse) and the pull-in winch. Two horizontal spools mounted on the locomotive's engine were powered by a crankshaft, which could be thrown into operation by a clutch. The gypsy locomotive's engineer could run his locomotive up to the log landing, throw the drivers into neutral, and activate the gypsy head on the pull-in winch, which would begin working like a donkey engine.[16]

With the coming of the donkey and the gypsy locomotive, log landings underwent a change in appearance. "Back landings" were built for tail holds. A back landing consisted of horizontal logs driven into the ground along the opposite side of the loading area and blocks chained at intervals along them. A line could be run from a donkey or gypsy locomotive through one of the blocks and directly across to a log, so as to pull the log onto a railroad car that had been pushed between the back landing and the skidroad.[17]

In logging the term "bull," placed in front of an endless number of other words, means big or strong. Around 1884 N. H. Pine, of Eureka, invented the bull donkey, capable of hauling logs over distances ranging from one thousand to two thousand feet. It was a portable machine, with an upright boiler in back and two large drums in front. Ahead of each drum were two upright and two horizontal spools, which slid back and forth on a spiral-geared shaft, feeding the line on and off the drum. The main drum belt held over 4,000 feet

of ¾-inch steel line. With the bull donkey, loggers could move logs down skidroads by steam power, and its advent signaled the end of oxen and horses in the woods.[18]

In August 1892, David Evans, E. H. Percy, and Bethune Perry received a patent for a skidroad equipped with snatch blocks that would keep a tow line in place on curves and turns. As many as twenty logs could be fastened to a first log and pulled in by heavy line. At the landing, a small Dolbeer donkey usually took over from the bull donkey and pulled the logs onto waiting flat cars.[19]

Donkeys have undergone many changes since Pine improved on Dolbeer's small machine, and the term now covers a great variety of steam, gas, diesel, or electric power plants, plus drums to hold wire rope, used to haul logs from the woods, load them at landings, move equipment, rig up trees, and — in the old days — lower cars down inclines. Donkeys are referred to in many ways: by their use, such as a skidder, loader, roader, slacker; by the number of their drums, such as two-drum or three-drum; by other features, such as simple, compound, two-speed; by the size of their cylinders (in steam donkeys), such as 6 x 10 loader, 10 x 16 roader; and by name, such as Flyer Duplex.[20]

By the early 1900's, local firms were turning out more and more donkeys for the redwood lumber companies. Numerous descriptions of these "monsters" appeared in local papers. The *Eureka Humboldt Standard* for September 14, 1907 reported:

> Humboldt County, widely known as a county famed for its gigantic redwoods, is not lacking in a concern capable of building machinery equal to the task of handling the monarchs and hauling the great loads of logs from the woods to points for shipment. A donkey, which is the largest portable bull ever built in the county, besides being larger than any imported machine, has just been completed by the Eureka Foundry Company in this city for H. S. Thompson of Bayside, and its massive size and novel construction is one to induce comment. A few of its numerous features will give an idea of its size and capabilities.
>
> It has double reversible 10 x 16 engines supplied by steam from a 60-horsepower boiler of the locomotive type, being 54 inches in diameter with over 80 2-inch tubes, seven feet in length, operating on 150 pounds of steam. The boiler is carried on a 20-inch I beam frame with the remainder of the machinery. Massive cast-iron brackets bolted to the I beam frame carry the crank shaft, intermediate shaft, haul-back and main or haul-in drums. On the intermediate shaft, there are three steel pinions which may

Engineer Tom Kennedy running the bull donkey at Camp 9 in the Hammond Lumber Company woods in the 1920's. — ARTHUR PETERSON COLLECTION

Tom Kennedy near the huge main drum of the Camp 9 bull donkey in the Hammond Lumber Company woods. — ARTHUR PETERSON COLLECTION (BELOW) A long train of logs being pulled to the landing by the bull donkey in the Excelsior woods near Freshwater in 1890. — A.W. ERICSON

be shifted in or out of gear at the will of the engineer, two of which serve to drive the main drum by engaging the massive gear wheels which are at either end of the drum and which serve as the flanges for the same, while the other may be shifted into mesh with the haul-back drum, located to the rear and above the main drum. Each flange gear of the main drum has a weight of over 3,000 pounds and a diameter of over 6 feet. This drum is capable of carrying between 1-1/2 and 1-1/3 miles of 1-1/8-inch steel cable. The drum proper is three feet in diameter and 3 feet 10 inches between flanges. The levers, four in number, consist of the reversing brake, double and single pinion shifters, and are conveniently arranged for the engineer, within easy reach of his throttle valve lever. The entire machine is mounted upon wooden skids, being bolted from the I beam frame to the same. The skids are 16 x 20-inch pine timbers, 30 feet in length and 7 feet 9 inches wide. The entire weight of the donkey is between 21 and 22 tons. It was yesterday loaded on a special car to be hauled to its place of duty in Mr. Thompson's logging camp.

The donkey is entirely a home product, having been designed and built at the company's own foundry and machine works in Eureka, and in size and power makes one of the machines used in the northern pineries look like a coffee mill.

Some big machines were equipped with two haulback drums and could carry three miles of 3/4-inch wire rope.

When bull donkeys were doing the hauling, side spools were installed on all curves in the skidroad. A signal system over the full length of the road was also necessary. Two copper wires constituted the signal line, and dry cell batteries supplied the power needed to set off a large electric bell at the donkey site. Crews could thus signal the donkey puncher from any point on the skidroad.[21] The ride side of the first log in a train — the side that came in contact with the skidroad — was sniped, or beveled, and then all the logs were coupled together by means of grabs. A grab was an over-sized hook with a chisel bit point with a ring and dee. Each hook was babbitted to the end of a piece of 3/4-inch or 7/8-inch wire rope, four to six feet in length. Sometimes chains, instead of wire rope, were used. Two grab hooks were set into the lead logs, which were usually the biggest, while only one was set into the smaller logs at the end of the train. A train might contain as many as twenty logs.[22]

Grabs and other rigging were taken back to the woods from the landing in the pig, a sled approximately sixteen feet long carved out of half a log, sniped at both ends, hollowed out in the center, and fitted with a hold at either end. An empty pig

The bull donkey and crew, shown above, at work in the California Barrel Company woods east of McKinleyville in the late 1920's. — A.W. ERICSON (LEFT) A huge bull donkey in the Hammond Lumber Company woods which was used to load logs on railroad cars in the 1920's. — OTTO AND MARY EMILY DICK — G.J. SPEIER COLLECTION

The "California" bull donkey, which was built in Eureka, was used in the Hammond Lumber Company woods in the 1920's. — ARTHUR PETERSON COLLECTION At the left, a company official poses with the bull donkey crew in the Hammond Lumber Company woods in the early 1920's. — OTTO AND MARY EMILY DICK — G.J. SPEIER COLLECTION

A bull donkey crew pause for a picture in the Little River Redwood Company woods in 1929. — GEORGE KNAB COLLECTION

was attached to each train before it started for the landing, and a man known as a zoogler was responsible for taking care of it, following the load, and spooling the haul-back line onto the side spools along the skidroad. After the logs reached the landing, the pig was loaded with the grabs and other rigging and "skinned back" to the woods on the haul-back line. Even after logs began to be roaded by bull donkey, the water slinger still had to travel ahead of every load, keeping the skids moist.[23]

During the bull donkey era, smaller spool donkeys were also used for yarding and loading. Their rigging included wire rope, ranging in size from 5/8-inch to 3/4-inch; a becket hook, which resembled a grab; and small blocks, grabs, mauls, and axes. As a rule, horses packed water to these spool donkeys.[24]

Logging railroads continued to replace animal teams in the Redwood Country, and soon all logs were being transported from the woods to the mills by rail. The principal logging railroads were the Oregon & Eureka Railroad and the Northern Division of the Northwestern Pacific Railroad. In 1909 the Oregon & Eureka Railroad was operating three logging trains. One engine could haul fifty to sixty 70,000-pound capacity flat cars.[25]

The most popular engine used on the main logging roads was a Baldwin locomotive, a steam engine on which piston rods connected the cylinders to the driving wheels. On steep and crooked logging roads, and particularly on temporary spur lines where speed was not an important factor, geared engines with short wheel bases and extra power, such as a Climax, a Heisler, or a Shay, were used. In these engines, power from the cylinders reached the driving wheels by means of differing gear arrangements.[26]

Since the railroads ran parallel to the coast and against the natural drainage of the area, they were extremely expensive to build, and rugged terrain often added to the difficulty of constructing them. Construction costs varied from $30,000 to $60,000 per mile. Mainline roads had to be laid out with light grades and easy curves, so they could handle the heavy trains.[27] As a rule, logging railroads were built to a convenient point at the mouth of a gulch, where a landing anywhere from two hundred to three hundred feet long had been constructed. A heavy stationary bull donkey, with a drum capacity of 1-1/2 to 2 miles of 1-1/8-inch hauling line and from 3 to 4 miles of 3/4-inch haul-back line was hauled to the landing. Main skidroads, together with branches, were run out from the landing to serve the area to be logged, which ranged from 500 to 1,500 acres. Logs were yarded along the branches to the main road, usually by

The zoogler rides the pig into the landing in the Freshwater woods of the Excelsior Lumber Company in 1890. A zoogler is a polite description for a logger who tended the skidroad pig, a log attached to the incoming turn of logs, like the caboose on a freight train. — CLARKE MUSEUM COLLECTION

Loggers with their tools at a railroad landing in Humboldt County woods at the turn of the century. — HUMBOLDT COUNTY HISTORICAL SOCIETY

A railroad landing in the Hammond Lumber Company woods in the 1920's. The logs were loaded on railroad cars and hauled to the dump at the Samoa mill on the peninsula. — OTTO AND MARY EMILY DICK — G.J. SPEIER COLLECTION

Building a railroad bridge over Maple Creek in the Hammond Lumber Company woods in 1935. The 1945 forest fire burned most of the railroad bridges in the Hammond woods. — ALFRED THOMA COLLECTION

two small side-spool donkeys. There they were coupled into trains of from fifteen to thirty logs and dragged to the landing by the bull donkey.[28]

In 1949 the Union Lumber Company converted its logging railroad to a heavy-duty truck road, and large diesel trucks began moving logs rapidly to its mill. Truck roads are much less expensive to construct than logging railroads, and trucks have now almost completely taken over from the latter. The Georgia-Pacific Corporation (now the Louisiana-Pacific Corporation) at Samoa discontinued its mainline haul in 1960, and today trucks carry logs from the woods to the log dump. Throughout the 1960's and into the next decade, The Pacific Lumber Company continued to use a mainline haul from its Carlotta Camp to the mill, and during the four- to five-month dry season, trains also ran from its Larabee Camp to the mill at Scotia. The rolling stock belonged to the lumber company, but the train crews were supplied by the Northwestern Pacific Railroad on a contract basis. In 1977, however, Pacific Lumber ceased all rail operations and turned to trucks. The Simpson Timber Company is still operating the first railroad in California, the Arcata & Mad River Railroad, using three diesel electrics on a 7.5-mile mainline run from its mill at Korbel to the interchange at Korblex.

Chopping in an area to be logged took place anywhere from several months to a year before the actual hauling began, and all timber was felled at one chopping. Following the choppers, peelers removed as much of the bark from the trees as possible. Costly and dangerous, bark removal was for many years the redwood loggers' number one headache. Redwood bark is thick, stringy, and tough. If left on the logs, it will cause trouble in the production of good lumber at the sawmill. Left in the woods, it becomes hazardous. In 1928 The Pacific Lumber Company set up a large debarking

A winter log storage on Pudding Creek in 1910 for the Union Lumber Company. The creek was dammed in 1906 to hold 20 million board feet. — CARPENTER PHOTO — ROBERT J. LEE COLLECTION

An early picture of the Union Lumber Company at Fort Bragg. — ROBERT J. LEE COLLECTION

plant at Scotia, where bark was fed into a shredding machine. The end result was an insulation product. Since the market for such material was limited at the time, other lumber companies failed to follow suit. Today, however, hydraulic and mechanical debarkers have replaced the peelers and solved the debarking problem in the Redwood Country.[29]

Redwood bark can vary from four to eighteen inches in thickness, and so much slash, such as tops, bark, and limbs, accumulated on the surrounding ground as the peelers worked that it became well-nigh impossible to move through the area. The timber was left lying until the slash was tinder dry, and then the area was burned over. These slash burns usually took place in May or June and again in the late fall. Losses due to merchantable timber being burned along with the slash were not heavy. The danger of a burn running out of control was slight, since the high ignition threshhold of redwood, due to its non-resinous nature, slowed the rate of flame spread. Fire would traverse cut land rapidly, but in most cases stop immediately when it struck standing timber. After the general burn, which took only a few hours, had consumed the slash, men were sent in to extinguish whatever small fires were still burning among the logs. Slash burns also greatly reduced the chances of fire in the woods after machinery had been moved in.[30]

All the fire-fighting equipment available to woodsmen prior to 1936 would have fitted in the back seat of an automobile. Protection against fire was a primary concern only during slash burns. Indeed, an inaccurate conception on the part of the state legislature that redwoods would not burn at all kept the Redwood Country exempt from state fire laws for many years. After 1936, however, selective cutting — a whole new concept of forest management — was developed. Slash burning was incompatible with this practice, since any fire in the woods might damage trees due to be felled in the future. Accordingly, lumbermen learned to be highly fire-conscious, and lumber companies added more and more fire prevention equipment to their logging camps' inventories. Woodsmen today are as wary of fire as forest rangers, and the methods and implements they use to cope with it are completely up-to-date.[31]

After the slash burn, buckers, or crosscutters, sawed the felled trees into logs. All of this work was done by hand. About twenty-five percent of the cut timber would be sawed into 16-foot lengths, and the remainder divided equally into lengths of 18, 20, 24, 32, and 40 feet. Moving behind the buckers, swampers built skidroads with the help of a donkey engine. Once the cut logs reached the landing at the end of the skidroad, they were

Two pictures of a dead mile-long incline in the Freshwater woods of The Pacific Lumber Company in the 1930's. The loaded cars were lowered by cable from a stationary engine at the top of the hill. At the bottom of the incline a locomotive coupled onto the loaded log cars and delivered them to the log pond. The empty cars were returned to the top of the incline for another load of logs. — HENRY L. SORENSEN COLLECTION

loaded onto railroad cars.[32]

As early as 1881, Dolbeer & Carson initiated a gravity system of moving logs down a gentle incline by constructing a railroad from tidewater up Jacoby Creek three miles to their timber, on a grade sufficient to allow loaded cars to reach tidewater on their own momentum. Horses returned the empties to the timber site.[33] At about the same time that portable donkeys and railroad spurs improved log-to-landing methods, inclines — a system of lowering logs down a very steep railroad grade — also came into use. Together, these developments rendered skidroads obsolete. Conversion of numerous bull donkeys to incline use prompted the manufacture of incline machines specifically designed for the purpose. Brakes on converted bull donkeys were manually operated. Those on incline machines were air-powered. On grades steeper than thirty-five percent, loads had to be bridled to keep logs being lowered from slipping over the ends of the cars. The signal system on inclines was the same as the one used on skidroads.[34]

On some inclines, loaded cars were counterbalanced by empties; on others, a lowering engine let down the loads and pulled up the empties. On a "live incline," the power plant itself was reefed up and down the hill, either carrying logs on cars or skidding them on the ties. In the latter case, the power car was called a Barney, Dudler, or Dudley.

A 1-⅜-inch cable was attached to the top and bottom of an incline and wound several times around a big drum on the donkey or the power car. The engine on the car turned the drum and moved the car itself, along with a string of logs behind. On a "dead incline," the cars were lowered by cable from a stationary engine. In Sessom's dead incline system, a line from a hoisting engine passed through a huge block on a specially-built block car to an anchor on the side of the track away from the lowering engine. Loaded railroad cars were coupled to the block car, which was then lowered downhill in the bight, or sag, of the wire.[35]

The gypsy locomotive first patented by Dolbeer in 1883 underwent a series of alterations over the succeeding twenty-five years. By 1910 the donkey locomotive was geared with two gypsies, or spools, and two drums that carried six or seven hundred feet of ¾-inch cable. Its reversing engines were connected to the main axle, which had thirty-or thirty-six-inch drivers, by a clutch gear or noiseless chain. In turn, the drivers were connected by side rods to another pair of drivers. The locie's wheel base was about eleven feet. A pinion or crank shaft in the engine connected with a spur gear, fifty inches in diameter, on a shaft on one end of which were a friction drum and spool. This fifty-inch gear meshed into another gear of about the same diameter, also on a shaft carrying a drum and spool, so that the two spools were on

opposite sides of the engine. Spur gears and drums made up a complete donkey.[36]

Gears or a chain clutched to the axle connected the engine to the drivers. The clutch lever was in the cab, next to the engine's reverse lever, near the throttle. With the spur gears disengaged and the clutch on the axle thrown in, the gypsy was a handy little locomotive for switching and handling trains, with a running speed of eight to twelve miles per hour. With the clutch on the main axle pushed out and the clutch on the crank shaft pushed in (activating the fifty-inch gears), the gypsy became a double-drum donkey for yarding and general logging. As a locomotive, it could haul empty cars to the landing. As a spool donkey, it could haul and load logs. As a locomotive again, it could carry the load to the log dump[37]

The first decade of the present century also saw the development of skyline or high lead logging, a system derived from ships' rigging. Introduced in the northwest in 1906, skyline logging was under way in the Redwood Country two years later, and enabled lumber companies to operate year-round, regardless of weather. The high lead system made use of the donkey engine, but in this case, the lines were rigged to a spar tree and extended into the timber. A spar tree had to be tall, straight, sturdy, topped, and trimmed of its leaves. Guy lines were attached to keep it steady under the terrific pressure of tons of flying logs. High lead blocks and lines were strung from the spar tree to a tail spar at the outer end of the logging operation. Along this high line ran the roller block, from which logs and choker shackles (cables and hooks that cinched the logs tight) dangled. Loading blocks could also

The versatile gypsy locomotive in the Hammond Lumber Company woods in the early 1920's. — OTTO AND MARY EMILY DICK — G.J. SPEIER COLLECTION

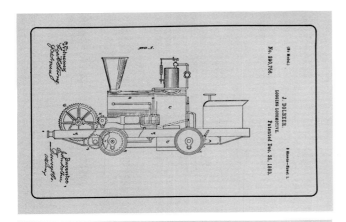

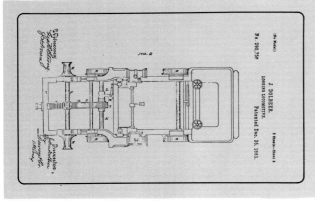

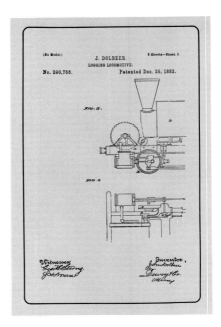

Drawings of the Logging Locomotive which John Dolbeer patented on December 25, 1883. — U.S. PATENT OFFICE

An early wood-burning "gypsy" locomotive in service in The Pacific Lumber Company woods. No. 2, a 2-4-2 tank engine, was built in 1887 by the Baldwin Locomotive Works. It was later renumbered to No. 22. — THE PACIFIC LUMBER COMPANY COLLECTION

An early Dolbeer locomotive in the Humboldt County woods. — CLARKE MUSEUM COLLECTION

be operated off a spar tree to load logs directly onto railroad cars and, later, onto trucks.[38]

Slackline logging, one of various skyline systems, became popular in the Redwood Country during the 1920's and 1930's. In slackline logging, a special machine known as a flyer lowered the mainline and skyline to the ground for taking on a turn (a log or string of logs), then brought the turn out of the woods in a single pull. Only the tail end of the turn was anchored, so the logs could be instantly elevated or lowered to allow for ground conditions, an advantage that eliminated much damage. For uphill logging to railroad spurs, the slackline method was highly efficient; but in downhill logging, it was extremely destructive.[39]

An accurate topographical map of the logging area, drawn to a scale of 300 feet to 1 inch, was the most important requirement in slackline logging. Landings had to be located in areas that allowed for maximum deflection, since the greater the deflection, the greater the load that could be carried without straining the rope. The distance from the landing to the long corners at the back end of the timber setting was usually less than 2,000 feet; but occasionally it would be greater, and when that happened, an extension was mollied, or spliced, to the end of the skyline.[40]

The essential equipment in slackline logging included 2,600 feet of 2-⅛-inch skyline; 3,000 feet of 1-⅜-inch mainline; 6,000 feet of 1-inch haul-back line; 6,000 feet of ½-inch straw line; a 2,460-pound skyline high lead block, with a 42-inch sheave; a 1,540-pound mainline block, with a 36-inch sheave; 5 haul-back blocks, with 16-inch and 18-inch sheaves; an 870-pound skyline carriage, with 21 6 x 3-½-inch sheaves; 16 guy lines, 2 inches and 1-½ inches in diameter; tree straps; and tail tree rigging. Total cost for this today would be in the neighbor-

A logger about to unload redwood logs with a jack screw, the important tool in the early days. — CLARKE MUSEUM COLLECTION (BELOW) The small gypsy locomotive at work in Humboldt Lumber Mill Company woods in the 1890's. — HUMBOLDT STATE UNIVERSITY LIBRARY COLLECTION

hood of $32,000. The slackline flyer, fully equipped and ready to log, weighed about 220 tons, and moving it by rail required a 16-wheel all-steel car. Slackline crews consisted of fourteen men, all highly skilled at their jobs.[41]

The Duplex Flyer, a donkey with two sets of engines in one used to power a double slackline system, was tried out in the Redwood Country, but proved unsuccessful there, although it was used extensively in the northwestern fir region. As one old-timer put it:

> Necessity was the mother of invention and we tried anything and everything to get the heavy logs out of the brush. Many of the methods of logging used in the northwest fir country never worked in the redwoods — the logs were just too heavy.[42]

In 1923 The Pacific Lumber Company tried using electric donkeys on its Freshwater area slackline operations, with limited success. The main objections to the electric machines were that they were hard on rigging and dependent on electric power in a region where storms, especially during winter months, frequently knocked down power lines[43]

From the standpoint of forestry and woods conservation, slackline logging turned out to be the

Loading an off-highway logging truck in the Big Lagoon operations of the Georgia-Pacific Corporation (now Louisiana-Pacific) in 1964. — DAVID SWANLUND

An early tractor at work in the Hammond Lumber Company woods in the late 1930's. — DAVID SWANLUND

most destructive logging procedure ever attempted in the Redwood Country. In areas logged by this method, not a single seed tree was left standing. One sorry example is the Freshwater area, owned by The Pacific Lumber Company. Seventy-five percent of the logging there was done by slackline, and the company is still trying to heal the scars inflicted by the iron beast, which left no trees for reproduction.[44]

The crawler tractor was first used in the redwood region in 1915. Running on endless treads, it could go into the woods after logs, just as bulls and horses had done. Initially, most lumbermen were skeptical about it, doubting that a tractor could handle the heavy redwood logs economically. However, crawler manufacturers improved their tractors each year, and in 1934 The Pacific Lumber Company began tractor logging in its Monument Creek operations. The following year, the Hammond Lumber Company was using crawlers successfully at its Van Duzen River site. Further improvements brought greater acceptance, and the diesel-powered crawler became a universal tool for yarding redwood. Tractor logging developed into one of the most popular methods of logging ever employed in the Redwood Country.[45]

The adoption of the tractor marked a watershed in the history of redwood logging. It made possible a system of cutting that would leave cut-over lands in productive condition, which in turn assured lumber companies of a permanent source of supply for their mills. This system was known as "selective cutting," and involved selecting specific trees to be cut in an area and preserving the remainder. The idea was not new to forestry in general; but it was an innovation in the Redwood Country. At the present time, selective cutting is practiced by all but one of the major logging companies. The single company still operating on the old clear cutting basis did give selective cutting a ten-year trial, only to find the method unsuited to its particular area, where winds are heavy and the shallow root systems of residual trees cannot withstand their force.

The steel wheels on the arches (trailers to be pulled behind tractors) of the first crawlers soon gave way to rubber tires, resulting in a tremendous savings on rigging and less damage to residual trees. When tractor logging began, crawlers commonly yarded distances of up to 2,000 feet. Today, with the advent of trucks and mobile loaders, they seldom yard over 1,000 feet. Although the majority of lumber companies switch to high lead logging during the winter months, particularly on steep slopes to avoid erosion, they still use tractors for most of their logging. Recently, however, a growing concern for ecology has caused these companies to look less favorably on the tractor, and they are turning to methods that are not so destructive of the terrain, e.g., the skyline is being used again.[46]

Chapter Notes

1. Lindsey, "Statement of Reminiscences," in Owen C. Coy, *The Humboldt Bay Region, 1850-1875* (Los Angeles: The California Historical Society, 1929), 287; Peter Rutledge, private interview, Eureka, California, June 5, 1956; Frank Fraser, private interview, Fortuna, California, October 22, 1960.
2. Rutledge, June 5, 1956; Fraser, October 22, 1960.
3. Rutledge, June 5, 1956; Fraser, October 22, 1960.
4. *The Humboldt Times*, 85th Anniversary Edition, E-2.
5. Rutledge, June 5, 1956; Fraser, October 22, 1960.
6. Peter Rutledge, private interview, Eureka, California, August 20, 1956.
7. Fraser, October 22, 1960.
8. Rutledge, August 20, 1956.
9. *History and Business Directory of Humboldt County,* Lillie E. Hamm, publisher (Eureka: *Eureka Humboldt Daily Standard*, 1890), 52.
10. Peter Rutledge, private interview, Eureka, California, November 5, 1956.
11. Walter F. McCulloch, *Woods Words: A Comprehensive Dictionary of Loggers Terms* (Oregon Historical Society and the Champoeg Press, 1958), 49.
12. *The Timberman* XXXIV-5 (March 1933): 9.
13. Rutledge, November 5, 1956.
14. Ibid.
15. Ibid.
16. McCulloch, *Woods Words,* 76.
17. Fraser, October 22, 1960.
18. *Wood and Iron* XIX-5 (May 1893): 532.
19. Ibid. XVIII-2 (August 1892): 86.
20. McCulloch, *Woods Words,* 49.
21. Ted Carlson, "Logging Yesterday," speech before the Redwood Region Logging Conference, Eureka, California, March 7, 1963.
22. Frank Fraser, private interview, Fortuna, California, November 22, 1960.
23. Ibid.
24. Ibid.
25. Dewey Dolf, "Rail Transportation," speech before the Redwood Region Logging Conference, Eureka, California, March 7, 1963.
26. Louis Sundquist, private interview, Blue Lake, California, June 10, 1955.
27. W. W. Peed, "Methods of Redwood Logging," *Eureka Humboldt Daily Standard*, August 10, 1909.
28. Ibid.
29. Emanuel Fritz, "The Changes I Have Seen," *The Humboldt Times*, The Redwood Logging Conference Silver Anniversary, March 7, 1963.
30. Peed, "Methods of Redwood Logging."
31. Fritz, "The Changes I Have Seen."
32. Fraser, November 22, 1960; Peed, "Methods of Redwood Logging."
33. Rutledge, November 5, 1956.
34. Frank Fraser, private interview, Fortuna, California, December 10, 1960.
35. McCulloch, *Woods Words,* 94, 166.
36. Robert T. Earle, "Gypsy Locomotives in Humboldt County," *Eureka Humboldt Daily Standard*, August 12, 1910.
37. Ibid.
38. Rutledge, August 20, 1956.
39. Ibid.; McCulloch, *Woods Words,* 169, 201.
40. Carlson, "Logging Yesterday."
41. Ibid.
42. Frank Fraser, private interview, Fortuna, California, December 10, 1960.
43. *The Humboldt Times*, February 9, 1923.
44. Carlson, "Logging Yesterday."
45. Fritz, "The Changes I Have Seen"; McCulloch, *Woods Words,* 40.
46. Fritz, "The Changes I Have Seen."

A pond man moving a log to the jack ladder at The Pacific Lumber Company sawmill at Scotia. — DAVID SWANLUND

Cutting at the Sawmill

Early Redwood Country sawmills were primitive and crudely equipped. The first saws used in them were the Mulay (or Muley) and the sash saw. Both were single saws, although the latter was eventually developed into a gang saw. The Mulay was held taut by an overhead spring pole. A wooden beam attached to a crank on either a water wheel or a steam engine worked the pole up and down, while guide blocks exerted a side pressure on the blade. The sash saw, more efficient than the Mulay, was a vertical saw six to eight feet long, with a broad blade. A sash running between side guides stretched it, and a crude carriage driven by ratchets ran the logs through.[1]

In 1853 the mills of the redwood region were producing more than any other area on the Pacific Coast. Most of the early ones were powered by steam, because water power was insufficient for cutting the huge logs. By this time, the mills had sash-gang saws, containing two or more blades in the sash, as well as Mulays. Circular saws were also used, alone or in horizontal gang arrangements, to turn out laths and shingles. All the lumber produced in these first mills was rough-finished and green.[2]

Redwood sawdust, which contained much water, posed a major problem. When attempts were made to use it for fuel, the crude boilers could not keep up a full head of steam, frequently forcing a mill to close for the day. In Eureka the wet sawdust was finally disposed of by hauling it away and dumping it into the gulches of what is now the city's downtown section. In time, however, the installation of better furnaces rendered redwood sawdust adequate to create enough steam to run a mill's saws.[3]

In 1861 sawmill headrigs (saw assemblies) were greatly improved by the installation of the double circular saw in place of vertical saws. In March of that year, the first double circular headrig, equipped with a cable feed carriage, was installed in a mill at Eureka. (However, logs still had to be sawed into bolts first by sash saws.) The two circular saws, one above the other, were fastened to the arbor with hubs. The top saw ran in the opposite direction from the lower, thus keeping the sawdust from filling the kerf, or width of the cut, and causing the saws to bind. The hubs limited the size of timber that could be sawed. As the cable wound up on its drum, the carriage moved into the saw, and a slab was cut. As the carriage moved back, another slab was cut. This forward and backward cutting doubled the output of lumber.[4]

The double circular saw also cut faster than its

A crane unloading logs into The Pacific Lumber Company pond at Scotia while pond men move logs away from the dumping area. — DAVID SWANLUND

The double circular saw cutting a log, and the "skidder" turning the log in a sawmill about 1890.

predecessors and soon took over from them. Later models of these steel saws were larger and more efficient. Their diameter increased over time from twelve inches to sixty and seventy-two inches. Logs no longer had to be sawed into bolts by sash saws. They could be handled directly by the headrig. At times, however, the Mulay would still be used to split them.[5]

In 1863 John Dolbeer, of the Bay Mill (later known as the Dolbeer & Carson Lumber Company), applied for a patent on a tallying machine that would inventory the number of board feet in the mill. The invention was reported in *The Humboldt Times* for May 16 of that year:

> The instrument . . . bids fair to dispense with pen and pencil in taking account of lumber, grain, or anything else that requires correct accounts to be kept. It is the size of an ordinary clock, and looks much like the face of the same. On the front is a hand, and on the dial figures are arranged the same as the clock. The manner of working it is the same as setting a clock by moving the hands. For instance: In taking account of the lumber, a man calls "sixteen feet," the hand on this instrument is immediately moved from twelve to sixteen, and back to twelve, which is only the work of an instant. So on he may go, calling off any number until he has reached one hundred thousand. Then the back of the instrument is thrown open and every number is added up, presenting a correct total of the whole amount. The total is given by three separate hands, pointing to the three respective sets of figures, each acting as a check upon the other. The machine can't lie. Mr. Dolbeer has applied for a patent, and we think there will be no trouble in his procuring it.

80

Evans' Third Saw at the Dolbeer & Carson Lumber Company in 1880. — PETER PALMQUIST
COLLECTION

In 1869 David Evans received a patent for Evans' Third Saw. The sawmill headrig could not cut logs as they came from the pond. They had first to be cut with a vertical saw, and Evans' invention was aimed at speeding up the process.[6] Evans' Third Saw hung on a horizontal arbor above the double circulars and cut down from the top of a log. It was parallel to the two lower saws, and it cut into the log four inches farther out. A fourth and smaller saw hung on a perpendicular arbor and cut horizontally into the log just at the bottom of the cut made by the third saw. The idea of running the third and fourth saws was to rabbet out a piece of wood extending from the top of the log to a point just below the arbor of the middle saw. The three large saws were usually sixty to sixty-four inches in diameter, and mills that used them could cut logs eight feet in width.[7]

By 1880 mills in the Redwood Country were equipped with circular headrigs, pony saws, edgers, trimmers, planers, picket machines, jointers, shingle machines, lath saws, and tongue and groove machines. Within another few years, cable-driven carriages were replaced by faster steam-driven carriages. The first of these was installed in Eureka in 1884. Saws were now pushed through a log, rather than pulled through.[8]

In October 1885, the first band saw installed on the Pacific Coast went into operation at the Dolbeer & Carson mill in Eureka. Its large wheels — one below the main floor of the mill and the other just above the clearance space for carriage and log —were at right angles to the carriage. A continuous steel saw formed a band around the two wheels, which held the saw in place by friction. The *Weekly Times-Telephone* for October 31, 1885 carried the story:

> For a week or more a force of men has been at work in Dolbeer & Carson's mill making the necessary arrangements and putting up machinery for the working bandsaws. The work has been under the supervision of Messrs. H. S. Marland and Charles Eldridge, who represent the firm of Stearns & Co., who are introducing the bandsaw to this coast.
>
> The saw is a 16-gauge saw, or, about $1/16$ of an inch in thickness and is run on two pulleys nine feet in diameter. A trial was made yesterday and gave entire satisfaction, the saw cutting the boards as small as one-quarter of an inch in thickness, turning them out as true as it is possible for any saw to do.
>
> After running awhile at slow speed the feed was increased and the saw worked equally as well. The lumber turned out was far superior to that made by a circular, or pony saw. The bandsaw is far ahead of the circular, owing to the fact that there is from 20 to 35 percent less waste in sawdust from the former than from the

The sawmill setter on the carriage in The Pacific Lumber Company sawmill at Scotia in the 1920's. The sawyer at the right controls the carriage and the head saw. — PAUL JADRO COLLECTION

(ABOVE) A monorail in The Pacific Lumber Company yard, in the 1920's, moving a load of green redwood lumber from the green chain onto trucks to be moved to the sticking platform. (BELOW) An early straddle truck with solid rubber tires in TPL Company yard in the late 1920's. — PAUL JADRO COLLECTION

The sawmill sawyer of the Arcata Redwood Company examines a cant cut from a redwood log at the mill in Orick. — DAVID SWANLUND

latter. It will be seen from this that a great saving of lumber will be made in the course of a year. There seems to be no question as to the success of this saw, and it will probably not be long before many of them are in use in this country.

The whole weight of the machinery necessary for running is eleven tons, and the cost in San Francisco is $2,300. From here, Messrs. Marland and Eldridge will go to Fort Bragg, Mendocino County, where they will put in their second one on the coast, this one being the first they have put in.

Eventually the band saw — whose thinner blade produced less sawdust — superseded the circular saw as the main saw in the redwood mills. The ones installed at the Dolbeer & Carson mill and at the Union Lumber Company mill in Fort Bragg were not used as the mills' headrigs. The first band saw headrig was put into the Elk River mill in 1888. That band had a circumference of fifty-one feet and could cut a plank eight inches wide. It proved so successful that all the major mills on the Pacific Coast soon adopted band saws in order to increase their production.[9]

In 1891 the Dolbeer & Carson Lumber Company introduced a method of air-drying and curing redwood lumber.[10] Experiments had shown that the quality of redwood lumber increased when it was cured for about two years in the salt air of the coast. Large drying yards were laid out for this purpose. The great disadvantage, of course, was the length of time lumber was tied up in them. Earlier, in 1889, John Vance had been experimenting with a dry kiln, and by the turn of the century, the kiln used today, which dries lumber in approximately two weeks, had been introduced and was being extensively employed by the mills.[11]

Piling lumber by hand for air-drying in the Excelsior Redwood Company yard on Gunther Island in 1890. — A.W. ERICSON

During the late 1800's, many Redwood Country lumber companies employed from 300 to 400 men in their mill or mills during the busy season. Monthly wages ranged from $35 per month, paid to boys employed tying shingles, to $180 per month, paid to the head men. Other specific monthly rates were $30 to $50 for yard laborers, as well as helpers inside the mill; $50 to $85 for edgers and trimmers; $85 to $100 for sawyers and filers; $100 to $125 for engineers and machinists; and $75 for a talleyman. Transient workers received $3 per day.[12]

After the early 1900's, redwood sawmills did not change greatly. Procedures remain almost the same today, although machinery has been improved, and the new Louisiana-Pacific mill at Samoa is operated by an automatic push-button system. From 1921 on, Redwood Country sawmills began switching from steam power to electricity. The Pacific Lumber Company, getting the jump on its counterparts, dismantled its steam plant and electrified its mill A that year. Electrification increased the number of employees there by 150, to a force of 1,500 men working at one mill.[13] In 1922, finding its mill equipment outdated, the Dolbeer & Carson Lumber Company made plans to electrify the facility. Construction began immediately, and the new plant went into operation early in 1924.[14]

Stacking lumber for air-drying in the Simpson Timber Company yard in Arcata. — DAVID SWANLUND

Chapter Notes

1. Peter Rutledge, private interview, Eureka, California, November 5, 1956; Howard Brett Melendy, "One Hundred Years of the Redwood Lumber Industry, 1850-1950" (Ph.D. dissertation, Stanford University, 1952), 60.
2. *San Francisco Daily Alta California*, October 9, 1853.
3. *The Humboldt Times*, December 31, 1899; Melendy, "One Hundred Years," 60.
4. *The Humboldt Times*, March 30, 1861.
5. Peter Rutledge, private interview, Eureka, California, November 10, 1956.
6. Melendy, "One Hundred Years," 59.
7. *History and Business Directory of Humboldt County*, Lillie E. Hamm, publisher, Eureka; *Eureka Humboldt Daily Standard*, 1890), 47.
8. *The Humboldt Times*, February 11, 1882.
9. *Wood and Iron* IX-5 (May 1888): 71; Melendy, "One Hundred Years," 61-62; Thomas R. Cox, *Mills and Markets* (Seattle: University of Washington Press, 1974), 236.
10. Howard Brett Melendy, "Two Men and A Mill," in *Redwood Country: History, Language, and Folklore*, Lynwood Carranco, ed. (Dubuque, Iowa: Kendall/Hunt Publishing Company, 1971), 98.
11. Ralph T. Wattenburger, "The Redwood Lumber Industry on the Northern California Coast, 1850-1900" (Master's thesis, University of California, 1935), 37; Melendy, "Two Men and a Mill."
12. *History and Business Directory*, 55.
13. Rutledge, November 10, 1956; Melendy, "One Hundred Years," 205.
14. Rutledge, November 10, 1956; Melendy, "One Hundred Years," 205.

Stevedores loading lumber aboard the steam schooner *Yellowstone* in 1928. — NATIONAL MARITIME MUSEUM — SAN FRANCISCO

Shipping the Product

The first record of a shipment of lumber from Humboldt County to San Francisco appeared in that city's newspaper *Alta California* on April 28, 1851: "The brig *Cameo*, arriving from a voyage of exploration, had aboard a cargo of one hundred spiles from Trinidad Bay." Considerable coastal traffic between San Francisco and the Trinidad area had already occurred. The previous year, fourteen ships had sailed from San Francisco bound for Trinidad Bay because of the supposed proximity of the latter to the Trinity mines. Not long after, during the winter of 1850-1851, the discovery of gold in the beach sand about twenty miles north of Trinidad generated so much excitement that twenty-eight vessels, some of them steamers, headed in that direction carrying men eager to extract their fortunes from what they had named the "Gold Bluffs." Their enthusiasm waned once they discovered that the fine gold particles could not profitably be separated from the sand, and during the first half of 1852, only six vessels from the Humboldt region arrived in San Francisco. Later that year, however, lumber production around Humboldt Bay grew by leaps and bounds, opening a new era in shipping. By 1853 the lumber industry was firmly established along the northern California coast, where magnificent mountainous

forests ran down to the sea, and the only practicable method of exporting its products was by way of the Pacific Ocean.[1]

Humboldt Bay was not an easy harbor for ships to enter or leave. High seas and changing tides turned the approach to it into a raging storm of water that broke upon a treacherous sandbar, whose depth varied from ten to sixteen feet. Ryan, Duff & Company purchased the tug *Mary Ann*, to pilot sailing vessels in and out, in 1852. Nevertheless, by 1855 twelve ships had been wrecked in or near the entrance to the bay. The arrival of the *Mary Ann* coincided with a steady increase in the lumber trade. In 1853 no less than 143 ships left Humboldt Bay and arrived safely in San Francisco, each of them carrying from 45,000 to 200,000 board feet of lumber. Because the mills had overproduced, only 138 vessels made the trip the following year.[2]

Exportation of lumber to foreign ports began in 1854. In October of that year, Duff & Chamberlain shipped several cargoes to Australia. Later sailing vessels headed for Hawaii, Hong Kong, Chinese ports, the Sandwich Islands, and Chile. Although they were much larger than vessels used in the coastal trade, they were able to enter and leave the bay safely, and the profits more than justified this

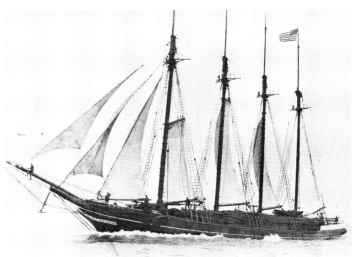

The sailing vessel, shown above, loads finished lumber at the Vance Mill & Lumber Company docks at Samoa in 1906. — NEIL PRICE COLLECTION (BELOW) The schooner *Esther Buhne* leaves Humboldt Bay with a load of lumber. — CLARKE MUSEUM COLLECTION

Sailing vessels loading at the Excelsior Redwood Company mill on Gunther Island in 1890. — NORTON STEENFOTT COLLECTION

Sailing vessels, shown above, on Humboldt Bay in the 1890's. — CLARKE MUSEUM COLLECTION (LEFT) The schooner *Eric* was a frequent visitor to Humboldt Bay at the turn of the century. — NEIL PRICE COLLECTION

extended trade. Meanwhile, ships of all shapes and sizes began venturing into the coastal trade, making stops at Big River and Crescent City, as well as Humboldt Bay.[3]

The period from 1855 to 1860 witnessed a decrease in shipping resulting from a general financial depression along the Pacific coast. In July, August, and September of 1858, almost all lumber transport ceased when gold was discovered on the Fraser River in British Columbia and goldseekers took over all available vessels.[4]

By 1863 the lumber trade was again on the upswing, and nine mills were functioning in Humboldt County, seven operated by steam and two by water power. The increased activity stimulated the lumber-carrying trade. Hitherto, San Francisco had been the chief domestic destination, but in 1861 lumber-carrying ships also headed for Sacramento (by way of San Pablo Bay and the Sacramento River), San Luis Obispo, and San Pedro in Southern California, and schooners carried several cargoes to Tahiti in the South Seas. Business was so good, in fact, that when the *Mary Ann* was lost, while towing the brig *Aeolus*, lumbermen promptly bought the tug *Merrimac* to replace her. Unfortunately, the *Merrimac* floundered after only a short period of service, and this second disaster depressed trade to such an extent that most of the mills closed down for a time, while others sharply cut back their production.[5]

Eventually, the number of ships leaving Humboldt Bay rose again. From April 1, 1867 to April 1, 1868, 318 vessels crossed the hazardous bar. For the succeeding twelve months, the figure was 394. For the entire calendar year of 1869, it was 424. This steady climb continued, and the total for the year 1876 stood at 1,106 ships, in spite of the fact that the lumber industries did not escape the nationwide Panic of 1873. During that economic crisis, all the mills had trouble finding markets for their lumber, and recovery in the Redwood Country lagged two or three years behind the national pattern.[6]

The following excerpt from *The Humboldt Times* for October 26, 1881 gives a picture of the amount of shipping activity to be seen on Humboldt Bay at the time.

> There were twenty-five vessels in port. The schooners *Ida McKay, Lottie Carson, W. H. Stevens* and *Fairy Queen* were loaded and ready for sea. The schooner *Jennie Thelin* was discharging ballast at Baird's Wharf, and the *Western Home* and *Ivanhoe* were loading at the Occidental Mill. The schooners *Emma* and *Louisa* are loading at Hookton (south bay), and the schooners *James Townsend* and *Bonanza* at the Arcata Wharf. At Carson's Mill the schooners *Sparkling Sea, John Hancock, A. P. Jordan,* and *Halcyon* were taking cargo. The *Mary Buhne* was discharging coal at Buhne's Wharf. The brig *Josephine* and the schooners *Sparrow* and *Mary Swan* were loading at Vance's Mill; the schooners *Mary E. Russ* and *Serena Thayer* at Russ & Co.'s Mill; and the schooners *Jessie Nickerson, Eva,* and *Isabel,* and brig *Hesperian* at Jones & Co.'s Mill on the Island. The barkentine *C. L. Taylor* and the schooners *Laura Pike, Vanderbilt,* and *N. L. Drew* were outside yesterday, but the rough condition of the bar prevented bringing them into port.

Records show that 466 sailing vessels and 210 steam vessels, all loaded with lumber, left Humboldt Bay in 1888. For 1892 the figures are 483 sailing vessels and 282 steam vessels. The sailing vessels and half of the steamships carried lumber and forest products exclusively. Forty-five of these vessels, including thirty-six schooners, two barkentines, and seven steamers, or thirty-eight percent of the total number of ships engaged in the lumber trade, were owned or controlled by persons or companies in Humboldt County. Thirty-seven percent of departing cargo left in similarly owned ships.[7]

During the decade from 1865 to 1875, shipbuilding flourished on Humboldt Bay. The leaders in this industry were Euphronius Cousins and Hans Bendixsen. Cousins was a native of Hancock County, Maine. After serving an apprenticeship in the shipyards there, he established his own facility and sent many a seaworthy vessel down the ways. Deciding that oppportunity awaited him on the Pacific coast, Cousins came to California in 1865 and settled in Eureka, where he constructed the first shipyard on the bay (on land owned by William Carson). He operated the yard from 1871 to 1883, and among the vessels built there were the *May Queen, Western Belle, Joseph Russ, Mary E. Russ, Maggie Russ, Ruby Cousins, Lillebonne,* and *Hesperian.*

While he was still running the yard, Cousins, in partnership with Joseph Russ, built the Cousins Mill on Gunther Island and engaged in the manufacture of lumber, as well as ships. Later, after selling his share of the mill to David Evans, he joined Charley Heney and E. J. Dodge in organizing the Eel River Valley Lumber Company, constructing its Newburg Mill east of Fortuna and devoting himself to the lumber business for the next ten years. Around 1893 this enterprising Yankee left the Redwood Country for Aberdeen, Washington, where he built still another shipyard and remained actively engaged in running it until his death on June 9, 1901.[8]

Perhaps the foremost shipbuilder on Humboldt

The barkentine *Jane L. Stanford* ready to be launched at the Bendixsen shipyard at Fairhaven on the peninsula in 1892. — A.W. ERICSON COLLECTION

Hans Bendixsen, shown on the left, built 113 wooden vessels on Humboldt Bay from 1869 to 1901. (RIGHT) Captain Henry Cousins commanded many ships out of Humboldt Bay. In 1905 he organized the Humboldt Stevedore Company, and in 1906 he headed the Cousins Launch & Lighter Company. Captain Cousins served eight years as a member of the harbor commission before he died in 1922.

Shipyard workers at the Bendixsen shipyard pose before the steam schooner *Willamette* in her frame in the 1890's. — LOWELL F. MC DANNOLD COLLECTION

A steam schooner being built at the Bendixsen shipyard at Fairhaven. — NATIONAL MARITIME MUSEUM — SAN FRANCISCO

Workers building a steam schooner at the Bendixsen shipyard at Fairhaven in the 1890's. — NATIONAL MARITIME MUSEUM — SAN FRANCISCO

Bay was Hans Bendixsen, who was born on October 14, 1842 in Thisted, Jutland, Denmark. His apprenticeship in the shipbuilder's trade lasted four years, and he subsequently went to sea as a ship's carpenter. In 1863, after a voyage to Brazil, he traveled to San Francisco and took a job building ships there, later coming to Eureka, where he worked at Cousins' shipyard for two years and then started his own yard — which would later be known as the Mathews Shipyard — at the foot of L Street. The first vessel completed there was the *Fairy Queen*, a topmast schooner launched on June 26, 1869.

In 1872 Bendixsen moved his shipyard to Fairhaven on the peninsula, where he continued to build excellent ships that gained him a wide reputation. By 1875 his output had climbed to twenty-four vessels, including twenty schooners, two steamers, one brig, and one scow. Although Bendixsen occasionally ran into financial trouble, he always managed to overcome his problems and pay off his creditors. He sold the Fairhaven facility for $250,000 in 1901 and died on February 12, 1902. In his thirty-three years of building ships on Humboldt Bay, he launched 113 wooden vessels, 17 of them steam schooners.[9]

John Vance, who in 1855 purchased the Ridgeway and Flanders Mill at the foot of G Street in Eureka, and in 1874 built the Big Bonanza Mill at Essex, on the north bank of the Mad River, owned four vessels and had a one-sixth interest in a fifth. These ships were the *Sparrow*, built in 1869; the barkentine *Uncle John*, built in 1881; the *Challenge*, built in 1883; the *Oceania Vance*, built in 1887; and the *Lizzie Vance*, built in 1888.[10]

Between 1881 and 1912, by which later date it had sold all its vessels, the Dolbeer & Carson Lumber Company maintained a small fleet of its own for shipment of its lumber products.[11]

The name of Captain Charles Nelson appeared for the first time in the Redwood Country lumber industry in 1876, when John Kentfield sold him a one-ninth interest in two mills and the logging equipment of D. R. Jones & Company.[12] Nelson was born in Denmark on September 15, 1831 and came to California in 1850. By 1867 he was in the shipping business and associated with John Kentfield & Company, a firm that at the time conducted the largest retail lumber business in San Francisco.[13] In 1860 Kentfield and D. R. Jones had purchased the Smiley Mill at the foot of H Street in Eureka, which had been operating since 1854. In 1866 D. R. Jones & Company also built a new double mill on Gunther Island.[14] As these mills prospered, the number of ships they needed for hauling lumber cargoes increased, and Captain Nelson's Charles Nelson Company handled their

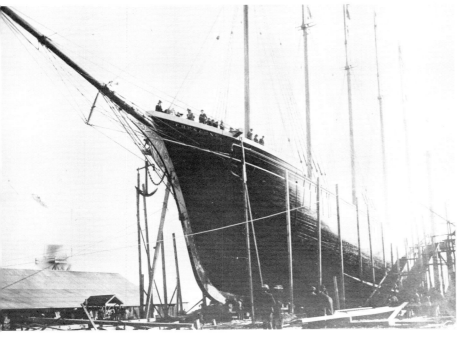

(ABOVE) Launching a steam schooner at the Bendixsen shipyard at Fairhaven. — A.W. ERICSON (LEFT) The *Crescent* ready to be launched at the Bendixsen shipyard in 1892. — CLARKE MUSEUM COLLECTION

Sailing ships loading at the Dolbeer & Carson Lumber Company wharves in 1901. The steamer *Corona* is at the right. — REED STROPE COLLECTION

Sailing vessels loading at the Arcata wharf in the 1890's. The vessel *Hayes,* out of San Francisco, is shown in the foreground. — A.W. ERICSON

Seven sailing ships loading at the Arcata & Mad River Railroad wharf in the 1880's. — CLARKE MUSEUM COLLECTION

The sailing vessel *Loudon Hill* loading at the Vance Mill & Lumber Company docks at Samoa about 1905. — GEORGIA-PACIFIC CORPORATION COLLECTION

Ships loading at the Arcata & Mad River Railroad Company wharf in the 1890's. The ships (from left to right) are the *Hampstead* (3,408 tons), the *Fulton* (380 tons), the *Foreric* (3,987 tons) and the *Lakne* (529 tons).
— HUMBOLDT COUNTY HISTORICAL SOCIETY

A sailing vessel loading at the Occidental mill in the early 1900's. The tug *Ranger* is heading north along the Eureka waterfront.
— CLARKE MUSEUM COLLECTION

Four sailing vessels and a steam schooner at the Vance Mill & Lumber Company docks at Samoa in the late 1890's. — CLARKE MUSEUM COLLECTION

shipping operations. In time he also secured interests in the Occidental Mill; Flanigan, Brosnan & Company; the Pacific Lumber Company; and the Humboldt Lumber Mill Company. Moreover, he was a major figure in the Northern Redwood Lumber Company at Korbel. Throughout the history of the redwood lumber industry, Nelson maintained his interests in various mills in order to assure his company of cargoes for its ships. In 1898 he owned twenty-three sailing vessels and four steamers, all engaged in the lumber trade.[15]

When Nelson's nephew, James Tyson, became its manager in 1904, the Charles Nelson Company was fast becoming known as the world's largest shipper of lumber. The Depression, however, coupled with the ever-present fluctuations and uncertainties of both lumber and shipping enterprises, did it in, and the company went into receivership during the 1930's.[16]

In 1863 surfaced redwood was selling for $26 per thousand board feet and rough redwood for $16.[17] In 1879 redwood prices dropped to their all-time low. Surfaced redwood sold for only $18 per thousand board feet and rough redwood for $12, and Humboldt County mills were operating at a reduced rate of production.[18] The year 1881, however, brought a big increase in orders from South America and Mexico, and by 1883 surfaced redwood had climbed to the price level of 1872, when it sold for $32 per thousand board feet.[19]

In 1887 the mills were again operating at full capacity. Orders poured in from San Francisco and other northern ports, and the first big real estate boom in Southern California created an even greater demand. As business expanded, more vessels were needed to keep pace. Between 1881 and 1893, fifty-six ships were built on Humboldt Bay.[20]

The Pacific Lumber Company of Scotia, which made its first shipment of lumber in 1887, from Fields Landing, gradually acquired ten vessels — owning some and holding a part interest in others — that took an important part in Humboldt Bay shipping. These were the *Helen N. Kimball*, built in 1881; the *Allen A* and *National City*, built in 1888; the *John A*, built in 1893; the *Aberdeen* and *Despatch*, built in 1899; the *Prentiss*, built in 1902; the *Temple E. Dorr* and *William H. Murphy*, built in 1907; and the steel vessel *Scotia*, built in 1919.[21] The Northwestern Pacific Railroad reached Eureka in 1914, and after that the lumber company found it difficult to keep all the ships profitably employed. It constructed both factory and storage facilities at Scotia, and began moving most of its lumber out by rail, although it continued to ship a small portion by sea.[22]

Another economic crisis, the Panic of 1893, sent

Five steam schooners at The Pacific Lumber Company docks at Fields Landing: (from left to right) the *Despatch*, the *Aberdeen*, the *Temple E. Dorr*, the *Prentiss*, and an unidentified vessel. — CLARKE MUSEUM COLLECTION.

The *Prentiss*, shown above, which was acquired by TPL in 1906, made 386 voyages for The Pacific Lumber Transportation Company. — THE PACIFIC LUMBER COMPANY COLLECTION (BELOW) The steamer *Scotia* crossing Humboldt Bar. — MELVIN A. KREI COLLECTION

The launching of the schooner *Ethel Zane* at the Peter Mathews shipyard on May 11, 1891. In the background is the Dolbeer & Carson Lumber Company and the Carson mansion. — CLARKE MUSEUM COLLECTION

Hans Henry Buhne

Hans H. Buhne, the foremost bar pilot on Humboldt Bay, was born in Denmark in 1822 and came to California in 1847 on a whaling ship. After some experiences in the mines, he was persuaded to ship as second mate on the schooner *Laura Virginia* which was bound for the mouth of the Trinity River. On their way north the crew discovered the entrance to Humboldt Bay. On April 9, 1850 (there are various dates), he took Captain Ottinger's boat and crossed the bar and entered the bay, the first American seaman to enter Humboldt Bay, and landed on the red bluff, now known as Buhne's Point. Four days later he brought in the vessel, and later he piloted the schooner over the bar and sailed to San Francisco. Soon after he arrived in the bay on May 6, 1850, and here he made a business of piloting ships in and out of the dangerous Humboldt Bar in a whaling boat. For the next 30 years he was the foremost bar pilot on the bay. By 1880 Buhne was part owner of the tugs *Mary Ann* and the *H.H. Buhne;* owned the bark *Watcher* which operated between San Francisco and the Orient; owned over 1,000 acres of farming and grazing land and 4,000 acres of redwood timberland; was part owner in four different vessels and one-third owner with D.R. Jones & Company in two sawmills; had a one-sixth interest in the Freshwater Logging Railroad; and had a two-thirds interest in the hardware business of H.H. Buhne & Company. Hans Henry Buhne died October 26, 1894. — DOROTHY BUHNE REDMOND COLLECTION

The *North Fork,* built by Bendixsen and owned by the Humboldt Lumber Mill Company and the Charles Nelson Company, leaves Humboldt Bay in the 1890's.
— PETER PALMQUIST COLLECTION

The *Newsboy* was the first steam schooner built for the redwood lumber trade. — NATIONAL MARITIME MUSEUM
— SAN FRANCISCO

the redwood industry into decline once again. Southern California's building boom collapsed, and freight rates to the East rose so high as to close the market.[23] Not until 1897 did markets gradually improve, and once more the redwood industry lagged behind others in climbing back to pre-panic levels of production.[24]

During the decade prior to the collapse of 1893, most of the lumber companies' foreign trade was with Sydney, Australia and Honolulu, Hawaii, and Humboldt County became the site of a U.S. Customs House in 1882.[25] Recognizing the importance of Humboldt Bay, the federal government not long after appropriated funds for improvement of the harbor, which consisted of dredging channels in the bay and constructing jetties at their entrances. The aim was to fix permanent direct channels of entry and remove the obstructing sandbar. Dredging began in 1887, and construction of the jetties started in 1889. By 1894 the initial work on the latter had been finished.[26]

Toward the end of the nineteenth century, steam schooners were replacing sailing schooners in the coastal lumber trade. As early as 1880, sailing vessels elsewhere had been equipped with small compound steam engines to provide them with auxiliary power. On the East Coast, such vessels proved successful in the whaling trade, and on the West Coast, similarly equipped ships carried perishable produce. Research has failed to reveal who first installed a steam engine on one of the small wooden ships used in the coastal trade; but thus converted, such vessels were tremendously improved. They no longer had to lie idle waiting for a favorable wind to propel them on their way. Also, auxiliary steam power gave them greater maneuverability in shallow harbors and against the challenges of the elements. When a sailing schooner was converted to a steam schooner, its outward appearance changed little. One slim stack rose from a box-like cabin erected on the afterdeck, just forward of the main mast. The vessel retained its sails, though sometimes in a modified form. Coastal operators watching the performance of sailing steam schooners were greatly impressed by them.[27]

Robert Dollar, owner of the Wonderly Mill at Usal, near Rockport on the Mendocino coast, ordered the first steam schooner with indigenous engines for use in the redwood trade. Built at the San Francisco shipyard of Boole & Beaton in 1888 and called the *Newsboy,* it was a small ship of only 218 tons that operated between San Francisco and Eureka. The original steam schooners were coal burners, and fuel oil was not used by the coastal lumber fleet until 1893. When they did switch to oil, however, operators altered the vessels by

The *Jewel,* which was built in 1888, was owned by the Caspar Lumber Company of Mendocino County. The vessel was wrecked at Caspar in 1899. — NANNIE M. ESCOLA COLLECTION

The *Klamath* being launched at the Bendixsen shipyard at Fairhaven in 1910. — MELVIN A. KREI COLLECTION

The *S.S. Fenwick,* owned by the Hammond Lumber Company, heads for the Humboldt Bar with a full load of lumber.
— CLARKE MUSEUM COLLECTION

The *Daisy Gray* leaving Humboldt Bay in 1950. This was the last wooden steamer to leave any lumber port in the Redwood Country. — MELVIN A. KREI COLLECTION

The *Tamalpais,* owned by A.B. Hammond, leaving Humboldt Bay with a load of lumber about 1927. — MELVIN A. KREI COLLECTION

The *Daisy Gray* loading at the Hammond Lumber Company docks at Samoa. — MELVIN A. KREI COLLECTION

In early 1917, James Rolph, Jr., the Mayor of San Francisco and later Governor of California, purchased part of the old Bendixsen shipyard at Fairhaven and part of another shipyard from the Hammond Lumber Company. The new company built many wooden ships, and the shipyard employed over 1,000 men. Fairhaven was renamed Rolph, and the town was a bustling community. When World War I ended and the demand for ships ceased, Rolph developed into a ghost town, inhabited mostly by commercial fishermen. The town was later renamed Fairhaven. — CLARKE MUSEUM COLLECTION

rounding their sterns, to make it easier for them to maneuver alongside wharves.[28]

After 1900 Southern California's demands for lumber greatly accelerated redwood production, and the coastal fleet was hard-pressed to handle the resulting increase in cargoes. To be sure their orders would be delivered, lumber companies began offering owners of steam schooners two- to four-year time charters, secured by surety bonds. In the face of this preference for steam-powered vessels, sailing schooners were forced out of the coastal lumber trade.[29]

Corporate ownership of vessels also became popular around the turn of the century. Previously, the ownership of ships had been shared by builders, chandlers, masters, business associates, and friends. The greater the number of partners owning a vessel in common, the less each partner would lose if it were lost at sea. Under the new system, each vessel was incorporated as a single-ship company, and shares of stock in the company (rather than fractional interests in the ship itself) were sold. In case of shipwreck or collision, individual stockholders in the corporation were liable only to the extent of their investment.[30]

From 1906 to 1909, following the disastrous earthquake and fire in San Francisco, three dozen new steam schooners slid down the ways, and the entire fleet was kept busy delivering the lumber needed to rebuild the city. Once San Francisco's needs were fulfilled, however, shipbuilding along the northern California coast slackened into a dormancy that lasted until the outbreak of World War I, when yards were once again deluged with ship orders.[31]

A. B. Hammond purchased the Vance Mill & Lumber Company in 1900 and began operating a fleet of steam schooners and larger vessels that transported lumber from Oregon and Humboldt Bay to San Francisco and Los Angeles.[32] In 1901 in Portland, Oregon, Hammond met Charles McCormick for the first time. McCormick had had twelve years' experience in the lumber industry in Michigan, and Hammond hired him to be sales manager of the Hammond Lumber Company. In July 1903 McCormick left that concern and, in partnership with Sidney Hauptman, started running both a shingle mill in Eureka and a lumber brokerage firm in San Francisco, The following year, Captain Edward Johnsen of the Hammond fleet took McCormick to Bendixsen's shipyard in Fairhaven, where the former Michigan lumberman made a down payment of $15,000 on the steam schooner *Cascade*. McCormick subsequently acquired the St. Helena Shipbuilding Company in Oregon, and by 1925 he controlled five coastal lumber shipping firms, which he merged into the McCormick Steamship Company. Eventually, this company extended its operations to South America, and that marked the beginning of the Moore-

99

The first engine-powered ship to be built on Humboldt Bay was the **Bloomington,** which was built at the Hammond Lumber Company shipyard. — CURTIS GILLIS COLLECTION

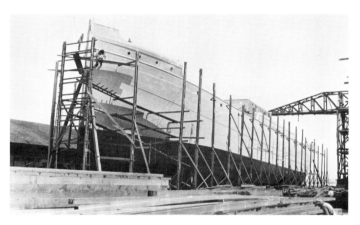

The Liberty ship *S.S. Keota* being constructed at the Hammond Lumber Company shipyard at Samoa in 1918. — NEIL PRICE COLLECTION

The *S.S. Keota* being launched at the Hammond Lumber Company shipyard May 2, 1918. — NEIL PRICE COLLECTION

McCormick Lines. In 1934 the McCormick Steamship Company was consolidated with the Pope & Talbot Lumber Company, and when — in 1940 —poor shipping conditions brought about the demise of the former, the Pope & Talbot Company took over the McCormick Steamship Company's assets. Charles McCormick died in Portland, Oregon, on February 24, 1955.[33]

The Hammond Lumber Company leased the Bendixsen shipyard at Fairhaven in 1907 and built thirteen ships there before giving up the lease in 1914. When the United States entered World War I, three years later, Hammond Lumber decided to build wooden ships at Samoa. Ways were constructed south of its mill, and the new shipyard employed 600 workers, who turned out seven liberty ships between 1917 and 1919. Paradoxically, although he was referred to as the "tycoon of the wooden fleet," Hammond was the first lumberman to use steel vessels for shipping lumber. By 1929 his company had sold off the last of its wooden ships.[34]

On August 30, 1929, the Sudden & Christenson Company joined the Hammond Lumber Company in establishing the Christenson-Hammond Lines, which operated a freight service that employed steamers belonging to several different owners. Its fleet consisted of six Hammond Lumber Company vessels, four Nelson Steamship Company vessels, two Sudden & Christenson vessels, and one McCormick Steamship Company vessel. In 1930 the new firm hauled 2,075 tons of freight.[35]

The Hammond Shipping Company, Ltd. was incorporated May 13, 1931. All its ten shares of stock were issued to the Hammond Lumber Company, whose fleet at this time was made up of the *S. S. Astoria, S. S. Covena, S. S. Redwood, S. S. Missoula, S. S. Samoa, S. S. San Pedro, S. S. Tillamook, S. S. Watsonville, S. S. Brunswick, S. S. Arcata, S. S. Portland*, and a controlling interest

The Matson Navigation Company steamer *Ventura* leaving Humboldt Bay with a load of lumber loaded on its decks. — DAVID SWANLUND

in the *S. S. Dorothy Winternote.*[36]

On January 27, 1939, the Hammond Lumber Company sold its fleet to the Hammond Shipping Company, Ltd. By 1942 the latter had sold all the ships, with the exception of the *S. S. Astoria, S. S. Arcata,* and *S. S. Brunswick,* and was about to close down its offices, when World War II intervened. The government made Hammond Shipping an agent of the War Shipping Administration, and for the duration, the company operated twenty-eight ships, including the liberty ships *A. B. Hammond* and *California Redwood.* At war's end, the company turned these vessels over to the government, and by December 1949 it had been dissolved. In the course of its nineteen-year life, the Hammond Shipping Company, Ltd. turned a profit of $246,816.34[37]

With the start of the pulp mills, the value of cargo leaving the Humboldt region rapidly increased. In 1950 the dollar value of Humboldt Bay's water-borne cargo was $4,752,000. In 1965, the year that the Georgia-Pacific Pulp Corporation came into being, the figure rose to $29,202,580. In 1967, with the Crown Simpson Pulp Company in operation as well, water-borne cargo value was $80,161,300. In 1968 it was $159,430,370. Although the figures for the following year show a drop to $125,550,200, Humboldt Bay remains today a major West Coast harbor.[38]

Les Westfall

Les Westfall became manager of the Humboldt Stevedore Company in 1952. In January 1962, Les resigned and started the Westfall Stevedore Company, and in 1971 he acquired the old Humboldt Stevedore Company which had been incorporated in 1906. Les now controls both companies. "At the present time," Les says, "the number of foreign cargo vessels loading on Humboldt Bay each year averages about 135, the average stay for a ship is three days, and about 2,700 seamen visit Humboldt Bay each year."

Six booms are working on the Matson Navigation Company's *Ventura* at Humboldt Bay as she loads redwood for the South Pacific. (LEFT) Stern boom of the *Ventura* with a load of heavy beams. Note how the lumber is chained to the upper deck. — BOTH DAVID SWANLUND

Chapter Notes

1. Owen C. Coy, *The Humboldt Bay Region, 1850-1875* (Los Angeles: The California State Historical Society, 1929), 121.
2. Wallace W. Elliott, *History of Humboldt County, California* (San Francisco: W. W. Elliott & Co., 1881), 129.
3. *The Humboldt Times*, October 21, 1854.
4. Coy, *Humboldt Bay Region*, 123.
5. Ibid.; Howard Brett Melendy, "One Hundred Years of the Redwood Lumber Industry, 1850-1950" (Ph.D. dissertation, Stanford University, 1952), 279.
6. Coy, *Humboldt Bay Region*, 267; Melendy, "One Hundred Years," 267.
7. J. M. Eddy, *In the Redwood's Realm* (San Francisco: D. S. Stanley & Co., 1893), 47.
8. Coy, *Humboldt Bay Region*, 263-264; Leigh H. Irvine, *History of Humboldt County, California* (Los Angeles: Historic Record Company, 1915), 808-809.
9. Irvine, *History of Humboldt County*, 809-810.
10. *The Humboldt Times*, April 27, 1878.
11. Melendy, "One Hundred Years," 193.
12. Ibid., 158.
13. *Wood and Iron* XLI-3 (April 1904): 10.
14. Jack McNairn and Jerry MacMullen, *Ships of the Redwood Coast* (Stanford: Stanford University Press, 1945), 66.
15. *Wood and Iron* XXXII-6 (July 1898): 12.
16. McNairn and MacMullen *Ships of the Redwood Coast*, 66.
17. Melendy, "One Hundred Years," 267.
18. *Wood and Iron* XXXII-6 (December 1889): 201.
19. *The Humboldt Times*, August 18, 1887; *Wood and Iron* XXXII-6 (December 1889): 209.
20. *The Humboldt Times*, August 18, 1887; Eddy, *In the Redwood's Realm*, 47.
21. Andrew Genzoli and Wallace Martin, *Redwood Pioneer, A Frontier Remembered* (Eureka: Schooner Features, 1972), 65-75.
22. Ibid.
23. *Wood and Iron* XIX-5 (May 1893): 517; Melendy, "One Hundred Years," 285.
24. Melendy, "One Hundred Years," 290.
25. Eddy, *In the Redwood's Realm*, 37.
26. *Souvenir of Humboldt County* (Eureka: Times Publishing Co., 1902), 38.
27. McNairn and MacMullen, *Ships of the Redwood Coast*, 14-17; Genzoli and Martin, *Redwood Pioneer*, 49-50.
28. McNairn and MacMullen, *Ships of the Redwood Coast*, 17-19.
29. Genzoli and Martin, *Redwood Pioneer*, 53.
30 Ibid.
31. Ibid.
32. *Hammond Redwood Log* IV-9 (September 1951): 1.
33. Edwin Coman and Helen M. Gibbs, *Time, Tide, and Timber: A Century of Pope & Talbot* (Stanford: Stanford University Press, 1949), 266, 268; Lowell S. Mengel II, "A History of the Samoa Division of Louisiana-Pacific Corporation and Its Predecessors 1853-1973" (Master's thesis, Humboldt State University, 1974), 44-45, 53-54.
34. Mengel, "History of the Samoa Division," 63, 73, 77, 78, 92, 94.
35. Ibid., 94-95.
36. Ibid., 113.
37. Ibid, 137, 141.
38. "North Coast Progress Report," *The Times-Standard*, February 27, 1970, E8.

At Noyo, ships were loaded from a platform hung from a wire cable, stretched from the cliffs out into the ocean and anchored on the vessel. The ship lay underneath the lower end of the cable, and the platform or sling let down by gravity feed was lowered when just above the deck. In this scene the schooner *Irene* is loading in 1916. Coming ashore are Captain Schuyler Colfax, his wife Bertha, the two children, a nurse, and a mate. — ROBERT WEINSTEIN COLLECTION

Logging the Mendocino Coast

The first fifty years of logging along the Mendocino coast constitute the most striking and colorful chapter in the history of the redwood industry. The exciting and hazardous logging operations common to all parts of the Redwood Country were accompanied there by an unusual and precarious method of loading lumber onto ships. The rugged Mendocino coast is eighty-five miles long, and the coastal mountain range runs close to the ocean along its full length. There is no coastal plain there, only cliffs in which the pounding seas have eroded innumerable indentations known to early deep-water sailors as "dog-holes." Redwood mills, located on the coast from Usal on the north to the Gualala River on the south, were built on the headlands. Wherever an indentation offered enough protection to serve as a tiny harbor, lumbermen constructed chutes and wharves. Sailing schooners would then find their way — through fog and heavy seas — into these dog-holes to take on lumber, railway ties, firewood, and tanbark. Every dog-hole had its own unique dangers, and many a schooner captain ran back and forth between San Francisco and one specific dog-hole, with whose every sunken rock he was familiar, even in the dark. In one stormy night alone, November 10, 1865, ten schooners were lost along the Mendocino coast.[1]

At Little River in Mendocino County, Captain Thomas H. Peterson and his crew of skillful craftsmen built twenty lumber schooners and sold them as fast as they were constructed. At Navarro Charles Fletcher built six or eight such vessels, and several more took shape at other Mendocino region facilities. Nevertheless, these were not enough to serve the entire Mendocino coast, and shipyards in San Francisco, Oakland, Benicia, and Vallejo also turned out schooners for the purpose.[2]

The chute was the answer to the problem of loading lumber along the rugged Mendocino cliffs. By the 1880's there was "a mill in every gulch, and a chute or two at every nearby indentation that offered slight protection from the prevailing sea." During that decade, seventy-six landings existed between Bodega Head and Humboldt Bay.[3]

At Newport, above Fort Bragg, a gravity chute carried lumber from the land to the ship below. As Newport was exposed to the open sea, it had no wharf — only several moorings where vessels were made fast by lines. These lines contained enough slack to permit the schooner to move with

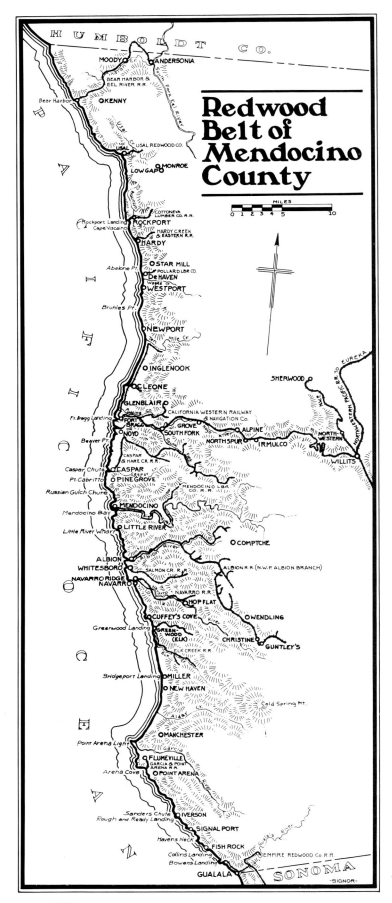

Redwood Belt of Mendocino County

MILES
0 1 2 3 4 5   10

Two schooners loading at Gualala by means of a wooden slide, called an apron chute. The sliding lumber was controlled at the ship end by means of an "apron" hinged to be raised and thus stop the material when it was just above the deck. A schooner could take on its cargo in two days. — NATIONAL MARITIME MUSEUM — SAN FRANCISCO

the rolling waves some twenty to twenty-five feet back or forward. The chute's apron projected slightly over the schooner's deck, and lumber was sent down its eighty-foot length one piece at a time. At the lower end of the chute was the "clapperman," who operated a brake-like device that slowed up each piece of wood and brought it to a stop just as it reached the apron. At that point, the ship's crew would haul the lumber off the chute and stow it. When the hold was full, they loaded it on the deck. As the vessel rolled, the distance between deck and chute would increase, and the waiting crewman risked injury or death if he failed to catch hold of a timber at just the right moment or the clapperman did not apply the brake in time.[4]

Captain Carl Rydell put in a three-month tour of duty on the steam schooner *Navarro* when a gravity chute was still operating at Navarro River:

> On my first day at Navarro we took on railway ties. Loading ties, which we called "sinkers," is particularly hard and dangerous work. If water-soaked, as they usually are, one of them is as much as a strong man can carry. I missed a sinker on a rough day in December under the chute at Navarro and left the Redwood Coast with a big toe in bad shape.

These steam schooners carried anywhere from 75,000 to 150,000 board feet of lumber, and along the Mendocino coast one can still hear tales of the skill with which they maneuvered among the rocks, tied up to moorings, and loaded with a full sea running.[5]

Steam schooners of the 1880's were better able to maneuver in cramped moorages than their predecessors, and this fact brought changes at the landings. Wooden chutes were gradually replaced

(ABOVE) Bowen's landing in 1897. From left to right, the *Newark* is loading bark under the wire; the *Mary Etta* is inside the harbor, and the steamer *Jewel* is outside; the *Barbara* is loading lumber, and the *Monterey* is outside. — NATIONAL MARITIME MUSEUM — SAN FRANCISCO (RIGHT) A vessel loading by wire chute at Bowen's landing. — NANNIE M. ESCOLA COLLECTION

by wire chutes, which allowed ships to load from points farther off shore. Wire chutes resembled the skylines used by loggers in the woods, and consisted of a heavy cable stretching from the ship to the shore and a carrier that traveled over it with the cargo to be loaded. The end of the heavy off-shore line was attached to a buoy. Once properly moored, a ship could pick up this off-shore line, while the in-shore line was brought out from the landing. Off-shore and in-shore lines were fastened to an assembly and attached to a boom on the ship. The main stress was against the off-shore anchor, and the cargo carrier traveled back and forth from ship to shore. Most wire chutes were operated by steam-powered drums at the shore end. This method of loading was used at such points along the Mendocino Coast as Greenwood, Albion, Little River, Caspar, Noyo, and Hardy Creek. The chutes themselves were built and rented out on a franchise basis.[6]

From 1860 to 1884, the heyday of the two-masted schooner, tremendous quantities of timber were cut, and an analysis of available records shows that over 300 schooners of the type that worked the Mendocino coast came out of the Pacific coast shipyards during that period. Although within a generation they were only a memory, the entire Mendocino region owes much to both sailing and steam schooners and the men who sailed them. Until 1912, when the Union Lumber Company extended its California Western Railroad from Fort Bragg on the coast to Willits in the interior, they provided the Mendocino County redwood industry's only link with the markets of the world.[7]

Steens Landing

Steens Landing was one of several ports along the Redwood Coast which used chutes for lumber loading in the 1880's. The view above shows how the lumber was lowered over the water and rocks to the schooner. (LEFT) A perfect photograph of the lower end of the chute showing the boom which lowered the chute to the vessel. — BOTH NANNIE M. ESCOLA COLLECTION

Iverson's Landing

Landings named for earlier settlers such as Iverson's were dogholes that could be challenged only by the smallest of ships and entered by the most superb of pilots. This picturesque spot was 91 nautical miles from San Francisco. — NATIONAL MARITIME MUSEUM — SAN FRANCISCO

Point Arena

The cape was first sighted by Ferrer in 1543 and called Cabo de Fortunas. The region appears on the maps of the following 250 years under various names. H.H. Bancroft in his book *History of California* listed the place as Punta de Arena. The Americanized version of the name is Point Arena. A town developed around a store built at the cape in 1859. Loggers found this shallow bay adequate for small lumber schooners. (UPPER LEFT) Lumber schooners and the wharf in a pre-turn of the century photograph. (LEFT) A close look at the lumber chute. (BELOW) An excellent view of the Point Arena doghole during the steamer era. Note the sturdier wharf and the shakes ready for shipment. The high bluffs are visible in the background. —

Greenwood

This cove was another along the coast that could be challenged only by the smallest of ships and entered by the most superb of pilots. The settlement was founded about 1862 at the mouth of the creek by Briton Greenwood, member of the second Donner relief party. (UPPER LEFT) A steam schooner leaving the landing in 1898. (LOWER LEFT) A small schooner loading off the wharf in calm water. — NATIONAL MARITIME MUSEUM — SAN FRANCISCO

Cuffy Cove Landing

The name of this small inlet, 111 miles from San Francisco, first appeared on the charts of the Coast Survey in 1870 as Cuffee's Cove, but it is not known whether Cuffee (or Cuffy) was actually a person's name or the nickname of a Negro settler. This photograph shows the three chutes at this landing. — NANNIE M. ESCOLA COLLECTION

Albion

Albion became one of the coast's better known lumber ports. The ancient name of Britain was applied to the river, as well as the land grant in 1844, by William A. Richardson, captain of the Port of San Francisco. Richardson, an Englishman, doubtless had in mind the name of New Albion, the name given Northern California by Drake. The town developed around the first lumber mill here in 1853. (LEFT) The curved wharf at Albion in 1897. — NANNIE M. ESCOLA COLLECTION

Building the three-masted power schooner *Sotoyome* on the Albion River in Mendocino County. The schooner was launched in December 1904. She was bound from the Coquille River to San Francisco when on December 7, 1907, about 14 miles northwest of Cape Mendocino, she caught fire from an explosion in the engine room. The crew was forced to abandon ship, and three hours later they were taken aboard the steamer *Lakme.* The steamer *Charles Nelson* put a hawser aboard the vessel and towed the burning ship and anchored it near the entrance to the Humboldt Bar. The *Sotoyome,* which was valued at $50,000, was a total loss. — WALLACE MARTIN COLLECTION

Little River

While cutting timber, Silas Coombs moved into Little River country, a cove with a tiny mountain stream. He immediately liked the land and set out a claim. In 1863 Coombs and friends built a mill at Little River and later set about building schooners of their own. Lumber was loaded by wire cable stretched from the land to an anchored vessel. Today the cove that once harbored lumber schooners now caters to abalone fishermen. (LEFT) Launching the *Electra* at the Thomas Peterson shipyard at Little River. — NATIONAL MARITIME MUSEUM - SAN FRANCISCO (RIGHT) Ships under construction at Peterson's shipyard. — NANNIE M. ESCOLA COLLECTION

Mendocino

The steamer *Yaquina* loading at the shipping point at Mendocino. — NANNIE M. ESCOLA COLLECTION It is said that the town was established as Big River, and later re-named Mendocino. Harry Meiggs, a promoter and operator of Meiggs Wharf, established his California Lumber Company on the bluffs overlooking the ocean.

Fort Bragg

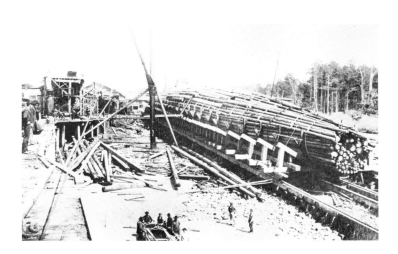

The town was named after the military post built to handle local Indian problems. It was established in 1857 by Lieutenant H.G. Gibson and named in honor of Lieutenant Colonel Braxton Bragg of Mexican War fame. Today Fort Bragg is the largest town between San Francisco and Eureka. About 1885, C.R. Johnson of Michigan migrated to the region and purchased the Stewart & Hunter Lumber Company. He bought the adjacent land, built a mill and established the Union Lumber Company. The first attempt at rafting logs on the high seas was made in 1884 from Nova Scotia to New York. Ten years later it was attempted with success with redwoods which float quite well. Rafts were built cigar-shaped and from 700 to 1,000 feet long, with a depth at the center of from 30 to 35 feet and a breadth of from 50 to 60 feet. The taper extends 100 feet from each end. In these views a cigar raft, or Robertson raft, is being assembled in a cradle at the Union Lumber Company at Fort Bragg in 1892. This raft was 336 feet long and 36 feet in diameter. On the way to San Francisco, the chains gave way, and most of the logs were lost. A small portion of the raft was towed to San Francisco, thus paving the way for similar shipments. (LEFT) The raft completely assembled at Fort Bragg. — MENDOCINO COUNTY HISTORICAL SOCIETY

Cleone Landing

According to Frederick W. Hodge, *Handbook of American Indians*, the name is derived from the name of *Keliopoma*, the northernmost branch of the Pomo Indians. The steamer *Cleone* preparing to load ties at the Cleone landing north of Fort Bragg. — ROBERT J. LEE AND THE MENDOCINO COUNTY HISTORICAL SOCIETY

Newport Cove

Newport chute was another doghole along the coast which was loaded from a chute. The schooner *Golden Gate,* which was built in San Francisco in 1874, loading at the Newport chute. — NANNIE M. ESCOLA COLLECTION

Westport Landing

This cove was a busy place despite the obvious handicaps of rocks. The age of steam brought more sophisticated machinery to the wharf as it changed from chutes to cable. — NANNIE M. ESCOLA COLLECTION

Rockport

A chute and a wharf were erected on the rocky coast line in 1876. A suspension bridge was built over the water to reach deep water cable loading. The track and wharf were three-quarters of a mile long, and the steel wire suspension bridge 275 feet long to the island. — NATIONAL MARITIME MUSEUM — SAN FRANCISCO

The schooner *Big River* approaches the San Francisco waterfront with a load of lumber from the Mendocino coast. — NANNIE M. ESCOLA COLLECTION

In 1867 a photographer caught two schooners riding out a storm, waiting to load up in the dogholes of the Mendocino coast. — NANNIE M. ESCOLA COLLECTION

Chapter Notes

1. Karl Kortum and Roger Olmsted, "A Dangerous-Looking Place — Sailing Days on the Redwood Coast," *California Historical Quarterly* XLX (March 1971): 49, 54.
2. David Warren Ryder, *Memories of the Mendocino Coast* (San Francisco, Taylor & Taylor, 1948), 24.
3. Ibid., 44; Jack McNairn and Jerry MacMullen, *Ships of the Redwood Coast* (Stanford: Stanford University Press, 1945), 21; Lynwood Carranco and John T. Labbe, *Logging the Redwoods* (Caldwell,
Idaho: The Caxton Printers, Ltd., 1975), 33.
4. Ryder, *Memories of the Mendocino Coast*, 28-29.
5. Kortum and Olmsted, "A Dangerous-Looking Place," 45; Ryder, *Memories of the Mendocino Coast*, 29.
6. Carranco and Labbe, *Logging the Redwoods*, 38; McNairn and MacMullen, *Ships of the Redwood Coast*, 25.
7. Kortum and Olmsted, "A Dangerous-Looking Place," 50.

Pacific Lumber Company No. 25 clatters across the Twin Creek trestle with a load of logs about 1907. This 2-4-2 tank engine was built by the Baldwin Locomotive Works in June of 1904 as No. 5. It was renumbered to No. 25. — ERNEST MARQUEZ COLLECTION

Redwood Country Railroads

Nowadays most of the lumber produced in the Humboldt Bay region moves out by rail. Approximately a hundred carloads leave the area daily. Any shipments of logs and lumber that depart by ship are usually bound for foreign ports. Like other elements of the lumber industry, Redwood Country railroads have a history.

After completion of a telegraph line between Eureka and Petaluma in 1873, residents of the isolated and mountainous Humboldt Bay area began pushing for a railroad to connect them physically with the outside world. Not until 1915, sixty-five years after the region was first settled, did such a line come into being. Railroads, however, had by then already contributed to the development of the region by way of the lumber industry. The first railroads in the State of California were constructed in Humboldt County in 1854 to haul logs to the water's edge, with horses or mules furnishing the motive power.[1]

On December 5, 1854, in order to create a port in the vicinity of Union (now Arcata), the Union Wharf and Plank Walk Company was incorporated for the purpose of building a wharf there that would extend out to a channel in Humboldt Bay. The following year, the wharf was completed, wooden rails were installed on it, and horses began hauling cargo over them. When steam came to the road in 1875, the company extended the railroad tracks overland to the Dolly Varden Mill, built in 1872 by Isaac Minor and Noah Falk. As other mills were constructed, the company kept extending the line to the northeast in order to serve them. In 1881 this railroad became the Arcata & Mad River Railroad, and today it is still serving the Simpson Timber Company's mill at Korbel.[2]

In 1874 John Vance, who had become the owner of the Ridgeway and Flanders mill at the foot of G Street in Eureka, constructed the Big Bonanza Mill at Essex, near his timber on the north bank of the Mad River. To transport his lumber, Vance built a railroad from the Big Bonanza Mill along the north side of the river, down to the bay at Mad River Slough. This new line was named the Humboldt and Mad River Railroad.[3] Vance's mill at the foot of G Street burned down in 1892, and that same year, his nephew and sons built a big mill across the bay on the peninsula. The Humboldt and Mad River was then extended from Mad River Slough down the bay side of the peninsula to the new mill. In 1893 the Eureka & Klamath River Railroad was incorporated to run from Eureka (by ferry) to Samoa, then to the town of Arcata, and on

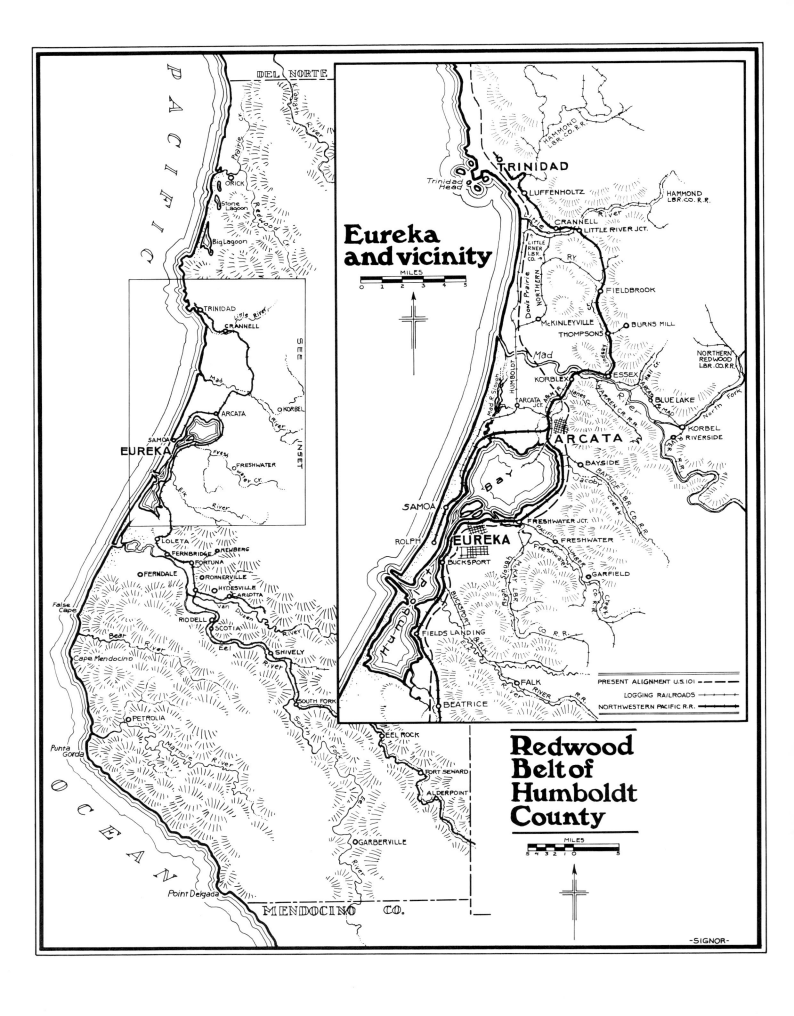

Eureka
and vicinity

MILES
0 1 2 3 4 5

Redwood
Belt of
Humboldt
County

MILES
5 4 3 2 1 0 5

PRESENT ALIGNMENT U.S.101 -----
LOGGING RAILROADS +—+—+—+
NORTHWESTERN PACIFIC R.R. ▪—▪—▪

PACIFIC

OCEAN

DEL NORTE

SEE INSET

MENDOCINO CO.

-SIGNOR-

ORICK
Stone Lagoon
Big Lagoon
TRINIDAD
CRANNELL
Little River
Mad
KORBEL
ARCATA
SAMOA
EUREKA
Freshwater Cr.
Elk River
LOLETA
FERNBRIDGE
NEWBERG
FORTUNA
FERNDALE
ROHNERVILLE
HYDESVILLE
CARLOTTA
RIO DELL
SCOTIA
SHIVELY
Van Duzen River
Eel River
SOUTH FORK
PETROLIA
Punta Gorda
Point Delgada
False Cape
Bear River
Cape Mendocino
Mattole River
EEL ROCK
FORT SEWARD
ALDERPOINT
GARBERVILLE
South Fork

Klamath River
Prairie Cr.
Redwood Cr.

TRINIDAD
Trinidad Head
LUFFENHOLTZ
CRANNELL
LITTLE RIVER JCT.
LITTLE RIVER LBR. CO.
Little River
HAMMOND LBR. CO. R.R.
NORTHERN RY
Dows Prairie
FIELDBROOK
McKINLEYVILLE
THOMPSONS
BURNS MILL
NORTHERN REDWOOD LBR. CO. R.R.
Mad River
KORBLEX
ESSEX
BLUE LAKE
Lindsay Cr.
ARCATA JCT.
WARREN CR. R.R.
North Fork Mad River
KORBEL
RIVERSIDE
ARCATA
Janes Cr.
BAYSIDE
BAYSIDE LBR. CO. R.R.
Jacoby Creek
SAMOA
Humboldt Bay
FRESHWATER JCT.
PACIFIC LUMBER CO. R.R.
FRESHWATER
ROLPH
EUREKA
BUCKSPORT
Freshwater Cr.
McKAY LBR. CO. R.R.
GARFIELD
Humboldt
Elk River
FIELDS LANDING
BUCKSPORT
FALK
BEATRICE

John Vance's Humboldt & Mad River Railroad locomotive bringing in a load of logs from the Lindsay Creek area. Note the "Gypsy" engine on the siding switching empties. — CLARKE MUSEUM COLLECTION

Large logs from the Big Lagoon woods are brought by railroad to the Vance Redwood Lumber Company log dump at Samoa. Note the sailing vessels loading at the dock at the left, in this 1900 circa illustration. — SAM SWANLUND

Officials of Vance Mill & Lumber Company pose before a large redwood log at the Big Bonanza mill in 1885. The locomotive *Onward* is at the left. — A.W. ERICSON PHOTO — DONALD DUKE COLLECTION

Loggers using screw jacks to unload logs from train cars into the log pond at Vance's Big Bonanza mill on Mad River. — ERNEST MARQUEZ COLLECTION

Three donkey engines are used to build extension No. 112 on the Eureka & Klamath River Railroad. Pilings 80 to 130 feet long are being driven into place by the Mercer-Hodgson Improvement Company, Eureka contractors and builders. — CLARKE MUSEUM COLLECTION

The "gypsy" locomotive of the Minor Mill & Lumber Company at work in the woods near Blue Lake in the late 1880's. In redwood country any locomotive which was a combination locomotive and donkey with two horizontal spools mounted on the engine where the pilot should be was called a "gypsy" engine. On such a locomotive the spools were powered by a crankshaft which could be thrown into operation by a clutch. It was a sight to see a railroad engine yarding logs with the hogger acting as a donkey puncher. The word "gypsy" should not be confused with the locomotive named *Gypsy* built by the Globe Iron Works of San Francisco for the Eureka & Klamath River Railroad. — DONALD DUKE COLLECTION

(ABOVE) Building a bridge over Balke Creek for the Little River Redwood Company. This was the highest single pole bridge in redwood country at the time. —A.W. ERICSON PHOTO — KATIE BOYLE COLLECTION (RIGHT) The Little River Redwood Company train crew, with a giant load of redwood logs in 1914, pose before Goat Rock near Bulwinkle (later Crannell). — BLILER PHOTO —KATE BOYLE COLLECTION

(LEFT) V. Zaruba, superintendent of the Humboldt Lumber Mill Company at Korbel, examines the Arcata wharf on his steam inspection car on June 1, 1893. — CLARKE MUSEUM COLLECTION (BELOW) An early photo of the Arcata & Mad River Railroad locomotive No. 5, an 0-4-0 tank engine carrying the name *Arcata,* built by Porter in 1882. — CALIFORNIA REDWOOD ASSOCIATION

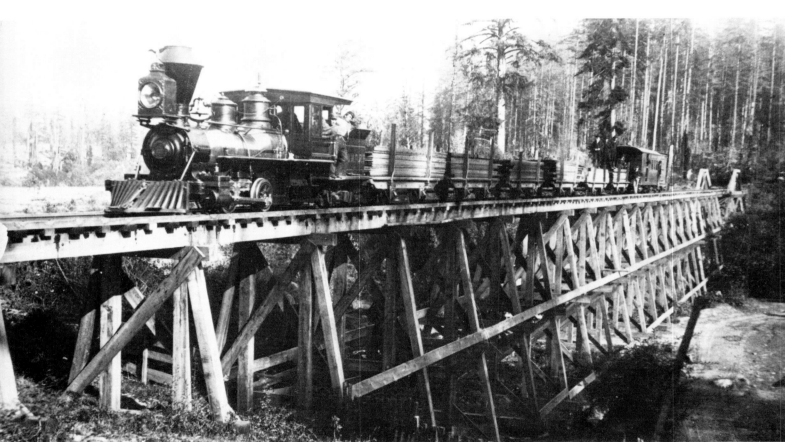

An 0-4-0 type locomotive of the Arcata & Mad River Railroad en route to the Arcata wharf. This train with six flatcars of cut lumber and a passenger coach/caboose crosses Warren Creek bridge. — PETER PALMQUIST COLLECTION

The *Advance* of the Arcata & Mad River Railroad bringing in loggers from the Vance woods in the Lindsay Creek area of Mad River in 1889. Alexander Christie, who came to Eureka from Ireland in 1888, is the second man standing on the right. — GILFILLAN PHOTO — SARAH MC CURDY COLLECTION

up to the new Vance mill. Four years later, in 1897, this fifteen-mile railroad was officially opened to the public. Later it was extended once again, this time north to Trinidad.[4]

In 1875 the South Bay Railroad and Land Company constructed five miles of track from the southern end of Humboldt Bay up Salmon Creek to serve the Milford Mill & Lumber Company, which had been developed by David Evans, John McKay, and H. A. Marks. After all the timber in the area had been cut, this railroad was transferred to Freshwater Slough and named the Humboldt Logging Railway. It later became the property of the Excelsior Redwood Company.[5]

The Hooper brothers formed the Trinidad Mill Company in 1869 and operated one of the early tram roads in Humboldt County. When this company was consolidated with the Dougherty and Smith mill in 1873, the tram road was converted to a standard steam railroad.[6]

The California Redwood Company, formed to buy up all the mills and timber in the Humboldt area, purchased, in 1883, the holdings of D. R. Jones & Company, the Cousins mill on Gunther Island, and the Hooper mill at Trinidad, along with 100,000 acres of redwood timber. This new company also controlled the Trinidad railroad mentioned above and the Humboldt Logging Railway. The following year, William Carson of the Dolbeer & Carson Lumber Company invested in the Bucksport & Elk River Railroad, which ran from Bucksport to the Dolbeer & Carson timber and to Falk, the site of the Elk River Mill & Lumber Company. This road, as well, came partially under California Redwood Company control. In 1886, when the California Redwood Company had to suspend operations because of fraudulent dealings in the acquisition of timber, the Excelsior Redwood Company was formed in its place, and it soon owned the Humboldt Logging Railway, which transported its logs from the Freshwater area to the bay.[7]

The Eel River & Eureka Railroad, built by John Vance, William Carson, and others in 1882, was an extension of the old south bay line. From Milford on Salmon Creek, it tunneled through Table Bluff to Loleta, and continued up the Eel River Valley to Alton and beyond to the Van Duzen River. That same year, the Pacific Lumber Company incorporated the Humboldt Bay & Eel

Four Humboldt Logging Railway trains of the Excelsior Redwood Company pose for A.W. Ericson, the famous photographer of the Redwood Country, at the Freshwater camp east of Eureka. The first two locomotives are each pulling 12 flatcars with 12 huge logs which contain a total of 136,804 board feet of lumber.

More redwood logs from the Freshwater woods roll toward the Excelsior mill on Gunther Island about 1890. The train is handled by No. 4, a 2-4-2 tank locomotive built by the Baldwin Locomotive Works. — CLARKE MUSEUM COLLECTION (LEFT) Loggers pose alongside a train of huge redwood logs before they are taken to the Excelsior mill. — DONALD DUKE COLLECTION

A tonnage train of redwood logs from the Excelsior Redwood Company's Freshwater camp on the way to Humboldt Bay. Several trainmen stand alongside one of the logs to show the enormity of this shipment. — A.W. ERICSON

The Hammond & Little River locomotive No. 6, a 2-4-2 tank engine, pauses for the wet-plate camera. This locomotive was built for the Southern Pacific by Baldwin in 1889. She became Dolbeer & Carson No. 1, Little River No. 6, and finally Hammond & Little River No. 6. — **CLARKE MUSEUM COLLECTION**

Carrying an empty tender, a Northern Redwood Lumber Company locomotive stops in the woods to wood-up. Slabs were passed to the man on the tender by hand. Note that the first three log cars carry logs split with powder. — CLARKE MUSEUM COLLECTION (LEFT) Engineer Archie Ambrosini works one of his last runs between Camp 9 and Korbel. — HENRY SORENSEN COLLECTION

(ABOVE) The Pacific Lumber
Company locomotives Nos. 29 and
26 at Shively, circa 1910. — THE
PACIFIC LUMBER COMPANY COLLECTION
(LEFT) Scotia residents have their
picture taken on two huge redwoods
brought in by a TPL locomotive in
the 1890's. — CALIFORNIA REDWOOD
ASSOCIATION

(LEFT) The Pacific Lumber Company's *Star* running in reverse, steaming through the redwoods. This 0-6-0 tank engine was built by Baldwin in 1875 for the Excelsior Redwood Company. (BELOW) Shay locomotive No. 26 in front of the Scotia mill with a train of logs cut from one tree in the Shady Run area in 1907. — BOTH THE PACIFIC LUMBER COMPANY COLLECTION

The Pacific Lumber Company locomotive No. 29 is working steam as it crosses the Van Duzen bridge. Engine No. 29, a 2-6-2 type, was acquired new from the Baldwin Locomotive Works in 1910. The train is returning to the loader with a string of empty cars. — THE PACIFIC LUMBER COMPANY COLLECTION

Oregon & Eureka Railroad No. 11 (Hammond Lumber Company) with a load of huge redwood logs from the Big Lagoon woods on its way to the Samoa mill on the peninsula. This locomotive was built in the Hammond Shops in 1911 from parts taken from an old Los Angeles County Railroad engine. — A.W. ERICSON

A Shay geared locomotive on a Hammond Lumber Company train crossing Freeman Creek bridge in the 1930's. The logs are headed for the mill at Samoa. — CLARKE MUSEUM COLLECTION

McKay & Company Railroad No. 1, an 0-4-0 engine with a "gypsy" engine on the front, brings in a load of logs from the Ryan Slough woods. The McKay tract is now owned by Louisiana-Pacific. — CLARKE MUSEUM COLLECTION

McKay & Company No. 2 with a load of logs from the Ryan Slough (Creek) area east of Eureka about 1905. The Occidental mill was operated by McKay for many years and was located on Humboldt Bay between A and B streets. — CLARKE MUSEUM COLLECTION

(ABOVE) A train in Willits, Mendocino County, with a huge load of lumber. (LEFT) Locomotive No. 4 with passenger train at Fort Bragg after the turn of the century. This Hinkley built 4-4-0 was acquired from the Southern Pacific in 1904 by the Fort Bragg Railroad and became California Western No. 4. — GERALD M. BEST COLLECTION

Engine No. 17 and a train of huge logs near Fort Bragg on June 13, 1935. Tank locomotive No. 17 was built by the Baldwin Locomotive Works. — GERALD M. BEST COLLECTION

California Western Nos. 45 and 46 on a freight loaded with lumber at Irmulco. No. 45, a 2-8-2, was acquired from the Medford Corporation, while No. 46, a 2-6-6-2 Mallet was obtained from Rayonier. — DONALD DUKE

California Western No. 52 switches at the Union Lumber Company at Fort Bragg at night. The 0-4-4-0 diesel switcher was acquired from Baldwin in 1949. — ROBERT HALE

Engine Nos. 1 and 2, of the Caspar, South Fork & Eastern Railway in the woods about 1885. No. 1, an 0-4-0T named *Jumbo,* was built by the Vulcan Iron Works in 1869. The Baldwin Locomotive Works built No. 2 as an 0-4-2T in 1885, and carried the name *Daisy.* — NANNIE M. ESCOLA COLLECTION

(ABOVE) Caspar, South Fork & Eastern No. 1, the *Jumbo,* on the Jug Handle trestle just north of Caspar. —GERALD M. BEST COLLECTION (RIGHT) A huge redwood dwarfs a youngster on the Caspar, South Fork and Eastern Railway. — REDWOOD EMPIRE ASSOCIATION COLLECTION

Locomotive No. 3, a 2-6-2 tank engine, named the *Smilax,* and No. 2, the 0-4-0T dummy, at the Caspar mill. — GERALD M. BEST COLLECTION

Caspar, South Fork & Eastern's No. 3 bringing in a train load of logs to the Caspar mill. — REDWOOD EMPIRE ASSOCIATION COLLECTION

Locomotive No. 5, a 2-6-6-2 Mallet named the *Trojan,* of the Caspar, South Fork & Eastern Railway, steams across the Jug Handle Creek trestle near Caspar. — GERALD M. BEST COLLECTION

Shay locomotive No. 2 of the Albion Lumber Company is about to get an assist from Northwestern Pacific No. 226, a Hinkley built 0-6-0, in the woods near Albion in 1908. — NANNIE M. ESCOLA COLLECTION

Shay No. 2 of the Albion Lumber Company. Engineer John Bertran and fireman Ed McDonald are on the left, and brakeman Poulsen is at far right. The Shay was built in March 1906 by the Lima Locomotive Works as No. 123 and later renumbered No. 2. She was scrapped in 1937. — NANNIE M. ESCOLA COLLECTION

North Pacific Coast No. 1, a 2-6-0 type narrow-gauge Baldwin, at the Camp 6 loop near Greenwood Creek in Mendocino County about 1904. — NATIONAL MARITIME MUSEUM — SAN FRANCISCO

Flanigan and Brosnan 2-6-2 tank locomotive at Bayside with a load of logs. The train is on its way to Humboldt Bay. The logs were then rafted to the mill in Eureka. — A.W. ERICSON

Moving a railroad camp to a new setting in the Hammond Lumber Company woods in the early 1930's. — KATIE BOYLE COLLECTION

A Hammond Lumber Company woods camp in the Big Lagoon woods in the early 1930's. — KATIE BOYLE COLLECTION

River Railroad, to be built from Fields Landing on the bay to the mill at Scotia; but under an agreement between the company and the Eel River & Eureka Railroad, the latter would handle the lumber trains that ran on it. In 1885 the Pacific Lumber Company constructed its own line from Alton Junction to the mill site at Scotia, and also extended a line south along the Eel River.[8]

These early railroads constituted an invaluable addition to the expanding lumber industry. They enabled lumber companies, logging at greater and greater distances from streams, to move away from tidewater and build mills close to their timber supplies. By 1887 fourteen mills were entirely or partially dependent on railroads, four on the Eel River & Eureka Railroad, two on the Humboldt and Mad River Railroad, and three on the Bucksport & Elk River Railroad.[9]

When A. B. Hammond came down from Oregon in 1900 and bought out the Vance holdings, they included the Eureka & Klamath Railroad. In an attempt to impede the expansion of the Santa Fe Railway in northwestern California, Henry E. Huntington, Vice-President of the Southern Pacific Company, as well as a major stockholder in the Hammond Lumber Company, purchased the Eureka & Klamath Railroad on May 18, 1903 for $1,150,000, and then leased it back to Hammond.[10]

Northwestern Pacific president Warren S. Palmer wields a silver spike maul as his daughter Alice steadies the gold spike that linked San Francisco with Eureka by rail at Cain Rock, October 23, 1914. — CLARKE MUSEUM COLLECTION (BELOW) The first passenger train from San Francisco reaches Cain Rock. — ANDREW GENZOLI COLLECTION

Throughout the early 1900's, the Santa Fe and the Southern Pacific battled for control of the lucrative redwood lumber traffic in the Humboldt Bay region. Santa Fe agents took over the Eel River & Eureka Railroad, on which the Southern Pacific had once held an option; the Pacific Lumber Company's road up the Eel River, past Scotia, to Shively; and the California & Northern Railroad, which ran north along the bay from Eureka to Arcata. The Southern Pacific, headed by E. H. Harriman, acquired the California & Northwestern Railroad, which in 1903 had extended a line from San Francisco Bay north to Willits.[11] Three years later, however, both Harriman of the Southern Pacific and Ripley of the Santa Fe decided there was not enough business in the area to sustain two major railroads, and on November 24, 1906, under a compromise agreement, they formed the Northwestern Pacific Railway Company, which began constructing a line south from Humboldt Bay and north from San Francisco Bay through the difficult mountainous region between. On October 23, 1914, at Cain Rock in southern Humboldt County, President Warren Palmer of the Southern Pacific drove in the last spike on the rail line that finally connected Eureka with Sausalito, although scheduled service over the new road did not commence until the following July. From then on, Humboldt lumber mills were no longer dependent on ocean shipping for transportation of their products. Later, in 1929, the Southern Pacific bought out the Santa Fe's interest in the Northwestern Pacific and assumed sole ownership of the road, which today remains a busy and efficient freight hauler, moving lumber out of the Redwood Country.[12]

Chapter Notes

1. Owen C. Coy, *The Humboldt Bay Region, 1850-1875* (Los Angeles: The California State Historical Society, 1929), 286-287.
2. Lynwood Carranco and Mrs. Eugene Fountain, "California's First Railroad: The Union Plank Walk, Rail Tracks, and Wharf Company Railroad," *Journal of the West* III (April 1964): 243-254.
3. Wallace W. Elliott, *History of Humboldt County, California* (San Francisco: W. W. Elliott & Co., 1881), 138; Lowell S. Mengel II, " A History of the Samoa Division of Louisiana-Pacific Corporation and Its Predecessors, 1853-1973" (Master's thesis, Humboldt State University, 1974), 12.
4. Lynwood Carranco and John T. Labbe, *Logging the Redwoods* (Caldwell, Idaho: The Caxton Printers, Ltd., 1975), 93.
5. Ibid., 93-94.
6. Ibid., 94.
7. Ibid., 103.
8. Gilbert H. Kneiss, *Redwood Railways* (Berkeley: Howell-North, 1956), 92-93; Carranco and Labbe, *Logging the Redwoods*, 112-113.
9. Howard Brett Melendy, "One Hundred Years of the Redwood Lumber Industry, 1850-1950" (Ph.D. dissertation, Stanford University, 1952), 176-177.
10. *Western Railroader* XXVI-1 (January 1963): 5; Mengel, "History of the Samoa Division," 52.
11. Donald Duke and Stan Kistler, *Santa Fe — Steel Rails Through California* (San Marino, California: Golden West Books, 1976), 55.
12. Kneiss, *Redwood Railways*, 133-135.

The world-famous Carson mansion in Eureka as photographed in 1902 by A.W. Ericson. This majestic home was built by William C. Carson in 1885-86, and remains today as one of the outstanding examples of Victorian architecture. — INGOMAR CLUB

Redwood Capitalists and Companies: The First Wave

Gold first lured men to the virgin wilds of the Redwood Country, and among the gold seekers were lumbermen from the eastern seaboard, also ready to make their fortunes by plying on the western coast of America the trade they and their forefathers had profitably engaged in on the shores of the Atlantic. Humboldt County, largest of the redwood-producing counties, became the backbone of this new lumber industry, although success did not come overnight to its new inhabitants. Most of the mills they built on Humboldt Bay during the early 1850's failed to prosper, because communication with their chief customers in San Francisco was hopelessly slow and unreliable. Moreover, the price of lumber fluctuated wildly, and many lumbermen were done in by the instability of the market.[1]

Ultimately, only a few of these transplanted Easterners reached pinnacles of financial wealth and power in their adopted environment. William Carson and John Dolbeer were two among them who climbed steadily to the top. Together, they overcame the formidable obstacles involved in moving logs more gigantic than any they had ever seen from woods to mill, acquiring or building equipment and machinery capable of handling the enormous redwoods, and finding suitable markets for their finished lumber.

William Carson was born in New Brunswick in 1825 and joined the Gold Rush to California in 1849. He and four companions arrived in San Francisco on April 1, 1850. From there they headed for the Trinity gold fields, where they worked in the mines through the summer and then, because food was scarce, decided to spend the winter around Humboldt Bay, where game was plentiful.[2]

Reaching Humboldt region, they learned that a settler named Martin White was getting ready to build a small sawmill that he estimated would turn out 4,000 board feet of lumber a day. Carson and his friends signed a contract to supply the necessary logs, and set up a logging camp between Ryan Slough and Freshwater. Spruce and other timber that a small mill could handle were available close to the slough. It was there that Carson and Jerry Whitmore cut the first tree ever felled in Humboldt County for a saw log.[3]

In March 1851, after logging through the winter, the four young men returned to the mines. This time, they stayed for a full year — until they learned that Ryan and Duff had constructed a sawmill on Humboldt Bay. Spurred by this news, they departed the Gold Rush once again and headed for the Sacramento Valley, where they purchased a team of oxen and drove it to Humboldt Bay. When they reached that destination, in

John Dolbeer

William Carson

August 1852, the party split up, and William Carson went into the lumber business.[4]

Two summers later, Carson leased the Mula (or Muley) Mill, at the foot of what is now I Street in Eureka, and started sawing. That fall he shipped — on the *Tigress* and the *Quoddy Belle* — the first all-redwood cargo sent out from Humboldt Bay.[5]

In 1863 Charles McLean, the partner of John Dolbeer, drowned when the tug *Merrimac*, on her way from San Francisco to Eureka, capsized at the entrance to Humboldt Bay. Carson purchased McLean's half interest in the Bay Mill, below the bluff of today's N Street in Eureka, and the following year, the Dolbeer & Carson Lumber Company came into being.[6]

John Dolbeer, born in Epsom, New Hampshire on March 23, 1827, had also arrived in California in 1850 with the gold seekers, and the following year, had joined the rush to the Gold Bluffs above Trinidad. He entered the fledgling Humboldt County lumber industry in 1853, when he joined Martin White, Isaac Upton, C. W. Long, and Dan Pickard in organizing the Bay Mill. The following year, a declining market brought about the collapse of the Humboldt Lumber Manufacturing Company, with which the Bay Mill, along with six other sawmills, was involved. In the ensuing financial crisis, Dolbeer managed to hang on to his interest in the mill, although all the other original partners lost their shares.[7]

Soon after they became partners, Dolbeer and Carson bought timberlands in the Elk River area, and in 1875 they purchased part of the heavily-forested Lindsay Creek land, north of Humboldt Bay in the Fieldbrook region.[8] In 1878, at a sheriff's sale, Carson bought the mill at Salmon Creek, near the southern end of the bay, for $23,650, and Dolbeer & Carson expanded its

A slab of clear redwood (2x70x12) at the Dolbeer & Carson Lumber Company in the early 1900's. — G.J. SPEIER COLLECTION

Residents of Fieldbrook spend Easter Sunday, March 30, 1902, in the Carson woods in the Lindsay Creek area. — G.J. SPEIER COLLECTION

The employees of the Dolbeer & Carson sawmill in 1904. John Zimmerman is third from right in forefront.
— MARY JANE PHEGLEY COLLECTION

lumbering activities. On February 11, 1879, the two men filed articles of incorporation for the Milford Land and Lumber Company. The Milford Mill, which would prove a sound investment, was equipped with a standard double circular head rig and could turn out 45,000 board feet per day. In 1884 Carson invested in the Bucksport & Elk River Railroad, a narrow-gauge line twelve miles long, which cost him $146,284 and would remain in use until 1950.[9] Dolbeer & Carson also had a logging claim three miles up Jacoby Creek during the 1880's[10]

The earliest surviving copy of articles of incorporation for the Dolbeer & Carson Lumber Company is dated 1886. At that time, a listing of the Board of Directors, together with the number of shares in the company held by each, read as follows: William Carson of Eureka, 998 shares; John Dolbeer of San Francisco, 998 shares; George D. Gray of Oakland, 2 shares; William Mugan of San Francisco, 1 share; and John Carson of Eureka, 1 share.[11]

After 1866 Dolbeer, who was in charge of the company's sales and marketing, made his home in San Francisco, then the major financial center of the West. William Carson, who supervised the logging and mills, remained in Eureka, where he built (in 1885 and 1886) his internationally famous Carson Mansion, one of the most outstanding examples of Victorian architecture in the United States. Standing at the head of Second Street, the redwood gingerbread manor has been attracting tourists these many years.[12]

An examination of the Dolbeer & Carson payroll books reveals that most of the company's employees came from Northern Ireland and the Maritime Provinces of Nova Scotia, New Brunswick, and Prince Edward Island. In 1890 William Carson voluntarily shortened the working day of the mills' crews from twelve hours to ten, without cutting wages, and in time other mill owners followed his example.[13]

By 1895 the company had logged out the timber and closed down the logging camp on Elk River, and its woodsmen moved into the Lindsay Creek holdings, which amounted to some 9,000 acres. The first trainload of logs from there arrived at Humboldt Bay in May 1897.[14] The Milford Land and Lumber Company continued producing lumber until 1902, when Dolbeer & Carson sold it, and the new owners moved its equipment to Fields Landing.[15] At the turn of the century, the Dolbeer & Carson Lumber Company held 16,270 acres of timberland, worth $1,093,128, and John Dolbeer's and William Carson's assets in the Bay Mill and affiliated operations totaled $1,342,628.[16]

Dolbeer died in San Francisco in August 1902, at the age of seventy-five.[17] Carson died in Eureka in February 1912, at the age of eighty-seven. As one of Eureka's most prominent citizens and leaders, he was active in and promoted many financial institutions and public utilities during his lifetime; but until two years before his death, he never ceased to devote the fullest personal attention to his lumber operations. With his passing, the Redwood Country lost the last of the handful of persistent and enterprising pioneers who turned long hard work into spectacular success.[18]

On December 16, 1950, The Pacific Lumber Company bought the Dolbeer & Carson Lumber Company for its timber holdings. In 1956 the mill, established in 1864 and the oldest redwood mill in the region, was sold to the Simpson Timber Company, to be converted into a fir plant. In 1960 the Park Loading Corporation of Portland, Oregon purchased the mill and the bay properties.[19]

Sailing vessels loading at the Dolbeer & Carson Lumber Company wharves about 1901. — CLARKE MUSEUM COLLECTION

John Vance

John M. Vance

John Vance's career in the lumber industry was a remarkable example of determination and foresight. Born in Nova Scotia, Vance was a ship's carpenter until he headed for California during the Gold Rush. He arrived in the Humboldt Bay region in February 1852.[20]

As a carpenter and millwright, Vance helped convert the steamer *Santa Clara* into the Ryan and Duff Mill, the first successful enterprise of its kind on Humboldt Bay. Later, after trying his hand at the mercantile business, he turned his attention to lumbering and, at a sheriff's sale in 1856, bought the mill at the foot of G Street in Eureka, built by Ridgeway and Flanders, who had failed in operating it. In March 1857, when an act of the California legislature deeded the entire waterfront to Eureka, the town decided to divide that property into lots for mill owners and sell it to them for a dollar per foot of frontage. Vance, who already owned much of the waterfront, paid the specified fee and received title to the west side of the wharf from F Street to H Street, which put him in control of all waterfront between F Street and J Street in Eureka.[21]

The *Merrimac* disaster in 1863 also claimed the life of Vance's partner, a Mr. Garwood of San Francisco, and the Nova Scotian continued their milling and logging enterprises on his own. In succeeding years, when lumber prices dropped so low that many companies went bankrupt, Vance retained his lumbering interests and wisely invested his money in timber tracts. As a result, he eventually became the owner of thousands of acres of the finest standing redwoods in Humboldt County.

John M. Vance came to Humboldt County from New Brunswick, Canada, in 1865, when he was 20 years old. Here he learned the millwright trade and became the boss of the mechanical department of the Dolbeer & Carson mill. Before John Vance died, he entrusted his nephew John M. with the management of his railroads, timber, and milling operations. When the Eureka mill burned down, John M. helped to build the huge Vance mill at Samoa. As his share of the Vance estate, John M. received the controlling interest in the Eel River & Eureka Railroad, which he sold to the Santa Fe Railway Company in 1903 for $700,000. John M. Vance died in Eureka on May 31, 1907.

John Vance's mill at the foot of G Street in Eureka in 1890. The mill burned in 1892, and a new mill was built on the peninsula. — DAN WALSH COLLECTION

After he had logged off all his timber behind Eureka, John Vance secured a right of way for a railroad that would run from Mad River Slough, at the north end of Humboldt Bay, to his redwoods on the Mad River. In 1875, in the heart of that fine timber, he built the Big Bonanza Mill on Lindsay Creek. The Big Bonanza could turn out 40,000 board feet of lumber a day. A Baldwin locomotive and thirty cars hauled lumber from the mill to upper Mad River Slough. From there the *Antelope*, Vance's stern wheeler, transported it to his wharf at Eureka. By this time, his two operations were providing work for 150 men.[23]

All the Humboldt County mills shut down in 1887, crippled by a nationwide economic crisis, and almost all woods operations also came to a halt. Vance, however, refused to stop logging. He felt a responsibility toward the employees who had been working for him for years. During the period of depression, he somehow managed to get food for them, and his men kept on logging until his 160-acre mill pond contained 8 million feet of timber. When orders began to pour in from Australia and Hawaii in the spring of 1888, they far exceeded the ability of the mills to fill them. Since many of the mills had no logs in stock, Vance booked orders for all of his and made a killing. With both mills running full blast, he stepped up his logging operations.

From the beginning, Vance took an active part in the development and advancement of the Redwood Country, and he led the way in the establishment of numerous industrial enterprises in the Humboldt area. At his death, in 1892, he was serving his second term as mayor of Eureka.[24]

Before he died, Vance chose his nephew, John M. Vance, to manage his railroad, timber, and sawmill operations, and also gave him the controlling stock in the Eel River & Eureka Railroad Company. When the Vance mill at the foot of G Street in Eureka burned down in 1892, John M. decided to rebuild across the bay on the peninsula, where there was plenty of cheap land available. The Vance Lumber Company bought its frontage area on the bay from the Samoa Land Improvement Company. John M. Vance dismantled the Big Bonanza Mill on the Mad River and moved it to Samoa on the peninsula, to be used temporarily, until the equipment for his new mill arrived.[25] He also extended the railroad down the bay side of the peninsula to the mill site. In 1895 Vance's Samoa Mill produced 60,000 board feet a day and employed 320 men.[26]

John Vance's log pond on Mad River for his Big Bonanza mill at Lindsay Creek. — A.W. ERICSON

A Fourth of July picnic at Vance's Big Bonanza mill on Mad River in 1875. — CLARKE MUSEUM COLLECTION

The new Vance mill at West Eureka (Samoa) in 1899, the largest mill in the county at that time.

David Evans

Another outstanding member of the redwood industry was David Evans. Born in Carmarthen-, sire, Wales, in 1838, he came to Humboldt County in 1857, when he was nineteen years old, and later became famous for his sawmill inventions, his ability as a millwright, and his skill as a timber cruiser. Ultimately, he procured so much timberland that his organization, the California Redwood Company, found itself in serious trouble with the United States Land Office. In the course of his lifetime, Evans was associated with the Occidental Mill, the J. Russ Company, the California Redwood Company, and the Excelsior Redwood Company. Popular and well-liked, he, too, served two terms as mayor of Eureka.[27]

Incorporated in San Francisco in 1883, the California Redwood Company promptly purchased the J. Russ & Company Mill, the D. R. Jones & Company holdings, the Hooper Mill at Trinidad, and 100,000 acres of redwood timber.[28] California Redwood, which included J. Russ & Company of Eureka (owned by Joseph Russ and David Evans) and Faulkner, Bell & Company of San Francisco, planned to obtain some of the best timberlands in Humboldt County and sell them to a Scottish syndicate. James D. Walker, of Faulkner, Bell & Company, traveled to Edinburgh, where he signed a contract with the syndicate under which he agreed to sell them 50,000 acres at $7 per acre. The syndicate promised to pay as fast as Walker secured the deeds.[29]

Soon afterward, the first big fraud that took place on the Pacific Coast under the Timber and Stone Act of June 3, 1878 began. When a dense area of redwood timber in northern Humboldt County was surveyed and opened to the public, the California Redwood Company, whose offices were in Eureka, immediately hired men to file on the entire tract.[30]

Mayor David Evans of Eureka, one of the owners of the Excelsior Redwood Company, gives the Humboldt County teachers a ride on the Excelsior train from Eureka to Freshwater in 1890. — A.W. ERICSON PHOTO — DAN WALSH COLLECTION

The employees of the Excelsior shingle mill at Freshwater in 1892. Top row: Charles Lindstrom, unknown, unknown, Jack Taylor, Monroe Whetstone, Ed Brodie. Middle row: all unknown. Front row: Charles Allord, Leland Whetstone, John Murray, Bulcholzer, unknown.
— HAZEL MULLIN COLLECTION

Joseph Russ

Joseph Russ had an interest in the Cousins mill and the Excelsior Redwood Company. Besides being a successful lumberman, Russ was the largest landowner in Humboldt County, owning more than 50,000 acres, which included 21 dairy ranches. He died on October 8, 1886, while serving his third term in the California legislature. — RICHARD HARVILLE COLLECTION

At the time, all a person had to do to take advantage of the provisions of the Timber and Stone Act was to go to the Land Office, file his claim, and prove he was a citizen of the United States or show his "first papers" to prove he had declared his intent to become a citizen, and take an oath to one or the other effect. Thereupon his entry would be allowed.[31]

The California Redwood Company ran men into the Land Office by the hundreds. In this way, the company obtained all the timberland as fast as it came on the market. As soon as it had secured the entire tract, it sent representatives to England, where they sold the entire body of land, located in a number of different townships, to the Scottish syndicate. By illegal means, California Redwood had secured about 64,000 acres of timber, worth $6,400,000. Pending the actual transfer of the timberland to the syndicate, the company began pulling out the patents to the varying claims at a cost of $25 each. In Washington, D.C., the General Land Office became aware of this abnormal rush for patents, and its commissioner ordered all as yet unpatented claims suspended and sent special agents to the Humboldt region to investigate.[32]

In Eureka the California Redwood Company bribed the first three special agents; but it was unable to buy off the fourth. As a result of that agent's report, the General Land Office immediate-

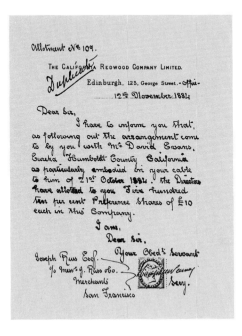

Correspondence between Joseph Russ and the California Redwood Company in Edinburgh, Scotland. — RICHARD HARVILLE COLLECTION

Employees in the Excelsior shingle mill at Freshwater. Those who can be identified are Ed Brodie (left), Charles Allord in rear, Leland Whetstone, and Monroe Whetstone. — HAZEL MULLIN COLLECTION

ly suspended between 150 and 200 entries and later cancelled them. California Redwood lost a small fortune in this unsuccessful venture. Although many of its principal representatives were indicted by a federal grand jury, David Evans was acquitted.[33]

The United States Government forced California Redwood to suspend operations in 1885. To hang on to whatever could be salvaged, David Evans and Joseph Russ, both of Eureka, and the Hooper brothers, Charles A. and George W., of San Francisco formed the Excelsior Redwood Company. They constituted its Board of Directors, and the new company filed articles of incorporation on April 10, 1886. Its main offices were in San Francisco.[34]

By 1888 the Hooper brothers represented the controlling interest in Excelsior Redwood, and they worked hard to secure markets. The "Excelsior" trademark became popular in all parts of the country, and the company's mill, on Gunther Island, ran day and night. The mill had excellent facilities for drying lumber and a good wharf from which to ship it. Two steam and five sailing vessels transported its products from Eureka to San Francisco. By 1893, however, Excelsior had logged all of the 24,000 acres of timber it owned, and the company soon ceased operations.[35]

****

The Sage Land and Improvement Company of Ithaca, New York was incorporated in 1893 and became established in Humboldt County in 1906, where it began buying large blocks of timber between the Klamath River and Prairie Creek. In 1916 the company bought 11,000 acres from the C. A. Smith Timber Company, and by 1924 Sage Land and Improvement, rated a $50 million concern, was the largest owner of redwood timber in the United States. In July 1956, the company sold its remaining redwood holdings to the Simpson Timber Company[36]

Gordon Manary, former general manager of The Pacific Lumber Company, examines a 16-foot diameter redwood log brought in from the Jordan Creek operations in 1944. — THE PACIFIC LUMBER COMPANY COLLECTION

The Pacific Lumber Company's original mill under construction at Forestville (Scotia) in 1885. The mill started operating in 1886, employing 150 men. — FRED ELLIOTT COLLECTION

The Pacific Lumber Company, which today owns the biggest redwood mill in the United States, came into being in 1869. Its directors were Alexander Macpherson and Henry Wetherbee of the Albion Lumber Company in Mendocino County, and although it kept control of the property it purchased until 1881, the company did not do any logging.[37] In 1882 Paxton and Curtins of Austin, Nevada and B. Low and James Rigby of San Francisco purchased Pacific Lumber and formed a new company, retaining the original name.[38]

This new Pacific Lumber Company started operating in 1886 at Forestville (later renamed Scotia), employing 150 men. The Eel River & Eureka Railroad was extended and used for moving timber to the bay shore at Fields Landing, where a huge wharf was constructed.[39] When fire destroyed the Pacific Lumber Company's mill in 1895, construction of a new and bigger one got under way immediately. Three years later, in 1898, Charles Nelson was elected president of the company.[40]

In December 1901, A. B. Hammond, who had recently purchased the Vance properties, bought forty percent of Pacific Lumber Company's minority stock, planning to take over the company. Unable to obtain more stock with which to do so, he subsequently sold out for $250,000.[41] The following year, H. C. Smith, a Minnesota capitalist and timber owner, purchased the majority stock of Pacific Lumber.[42] When the Pacific Lumber Company of New Jersey incorporated in 1903, it received all the property of the old Pacific Lumber Company of California. At the same time. H. C. Smith bought the Excelsior Redwood Company's holdings at Freshwater, which included mill, woods, camps, and railroad, and formed the Freshwater

The earliest photo of The Pacific Lumber Company at Scotia, taken in 1889. The stand of timber at the right was cut in 1907. — FRED ELLIOTT COLLECTION

The big mill of The Pacific Lumber Company at Scotia in 1902. — A.W. ERICSON

The Pacific Lumber Company at Scotia on the Eel River, thirty miles south of Eureka on Humboldt Bay. When The Pacific Lumber Company built their mill in 1885 the town was first called Forestville. The present name of Scotia was chosen on July 9, 1888 when the Post Office was established because many of the men working in the mill were natives of Nova Scotia. This is a current view of present day Scotia. — DAVID SWANLUND COLLECTION

150

Simon Jones Murphy   Albert Stanwood Murphy   Stanwood Albert Murphy

The Murphy family controlled TPL through the years. In 1905 Simon Jones Murphy (LEFT) gained control. His grandson, Albert Stanwood Murphy (CENTER), became president in 1931, and when he died in 1963, his son, Stanwood Albert Murphy (RIGHT), was named chairman of the board. Stanwood died in 1972. — THE PACIFIC LUMBER COMPANY COLLECTION

The beautiful setting of The Pacific Lumber Company and the town of Scotia on the Eel River. This community is a true frontier creation. The entire town dependent on a single firm for its existence, streets, homes, stores, gardens, school, bank, post office, museum, and library — all the appendages of one lumber company's mills. — DAVID SWANLUND

Edward M. Carpenter

In June 1933, after completing two years at Santa Rosa Junior College, Edward Carpenter found a summer job at TPL in Scotia. Here he learned the lumber business, and he became Assistant to the Manager in 1953, Resident Manager in 1957, Vice President of Operations in 1961, Executive Vice President in 1963, President in 1971, and finally Chairman of the Board and Chief Executive Officer in 1973. The Carpenter years were favorable ones for TPL's owners. The value of their shares increased almost sevenfold, and the potential market for their securities widened by having the shares listed on the New York Stock Exchange. — THE PACIFIC LUMBER COMPANY COLLECTION

Lumber Company. In 1904, in a shake-up of this organization, Smith resigned and E. M. Eddy, a major stockholder, became the new president.[43]

The Freshwater Lumber Company was discontinued the next year, and in 1905 the Pacific Lumber Company of New Jersey was succeeded by a consolidation of the Pacific Company, the Freshwater Lumber Company, and the Pacific Lumber Company. The directors of this hybrid, called The Pacific Lumber Company (TPL), were from Detroit and Saginaw, Michigan, and Chicago.[44]

Simon Jones Murphy soon controlled TPL, but died two months short of acquiring full ownership of the new company. Through a period of great demand for lumber, precipitated by the San Francisco earthquake in 1906 and accelerated by the outbreak of war in Europe in 1914, Murphy's sons and business associates operated and expanded TPL, acquiring additional timberlands and developing — in 1910 — the present company town of Scotia.[45]

Simon Murphy's grandson, Albert Stanwood Murphy, became president of TPL in 1931 and guided the company through the Depression. In 1940 TPL purchased large tracts of timber from the Dessert Redwood Company, the Hicks-Vaughn Redwood Company, the Hammond Redwood Company, and the Holmes-Eureka Lumber Company. Together, these purchases gave TPL a solid block of timber, twelve miles long and seven miles wide — 22,000 acres of virgin redwood — which the company claimed would be enough timber for twenty years of logging. The new acquisition was in the basin of Yager and Lawrence Creeks, behind Carlotta.[46]

Albert Stanwood Murphy, who had become Chairman of the Board of TPL in 1961, died in 1963. Stanwood A. Murphy, his son, was named Chairman on April 29, 1971. Stanwood Murphy died of a heart attack on August 2, 1972, at the age of fifty-three.[47]

In 1970 TPL bought twenty-one acres of industrial land in the south Fortuna area and constructed a high-speed sawmill at a cost estimated in excess of $1 million. This facility, which began operating in 1972, uses small logs thinned from TPL's young growth forests, as well as some of the smaller old growth logs that are not economically adaptable to manufacture in the company's mill at Scotia.[48]

TPL capital commitments at Scotia over the past decade have included multi-million-dollar facilities for plywood production, lumber storage and shipment, bark removal, and power generation. In keeping with the times, the company is also diversifying by buying agricultural property in the Sacramento Valley for production of field crops; acquiring timber-cutting rights to several hundred thousand acres of forest in the Rocky Mountains; and purchasing the assets of the Ideal Brush Company, which produces and markets brushes and other painters' supplies. Moreover, the Victor Equipment Company, a major manufacturer of gauges, valves, and welding equipment, is now merged with TPL. It is hardly surprising that in 1978 The Pacific Lumber Company reported the highest revenues and earnings of its 109 years of operation.[49]

Chapter Notes

1. *The Humboldt Times*, December 31, 1899.
2. Leigh H. Irvine, *History of Humboldt County, California* (Los Angeles: Historic Record Company, 1915), 607.
3. *The Humboldt Times*, December 31, 1899.
4. Irvine, *History of Humboldt County*, 608.
5. *The Humboldt Times*, December 31, 1899.
6. Ibid., August 4, 1904; Humboldt County Recorder, *Deeds*, Book D, 641; Howard Brett Melendy, "Two Men and a Mill," in Lynwood Carranco, ed., in *The Redwood Country: History, Language, and Folklore* (Dubuque, Iowa: Kendall/Hunt Publishing Company, 1971), 91.
7. *Wood and Iron*, XXXVIII-3 (September 1902): 18; *The Humboldt Times*, September 16, 1854; Melendy, "Two Men and a Mill."
8. Humboldt County Recorder, *Deeds*, Book F, 656 and Book I, 758 and Book P, 252-261; Melendy, "Two Men and a Mill"; *The Humboldt Times*, March 29, 1878.
9. *History and Business Directory of Humboldt County*, Lillie E. Hamm, publisher (Eureka: Daily Humboldt Standard, 1890), 69.
10. *Wood and Iron* VII-7 (July 1887): 69.
11. Humboldt County Clerk, *Articles of Incorporation*.
12. Melendy, "Two Men and A Mill"; Scoop Beal, *The Carson Mansion* (Eureka: Times Printing Company, 1973), 1.
13. *The Humboldt Times*, September 2, 1890.
14. *The Humboldt Times*, May 13, 1897.
15. *Wood and Iron*, XXXVIII-3 (September 1902): 1.
16. Melendy, "Two Men and a Mill."
17. *Wood and Iron* XXXVIII-3 (September 1902): 1.
18. Irvine, *History of Humboldt County*, 609.
19. *The Humboldt Times*, December 15, 1910 and January 14, 1962.
20. Irvine, *History of Humboldt County*, 1168.
21. Susie Fountain, "Early Days of Humboldt," *The Blue Lake Advocate*, November 15, 1956.
22. *The Humboldt Times*, August 5, 1904.
23. *The Humboldt Times*, February 7, 1878.
24. J. M. Guinn, *History of the State of California and Biographical Record of Coast Counties, California* (Chicago: The Chapman Publishing Company, 1904), 260.
25. *The Humboldt Times*, September 12, 1916.
26. *Wood and Iron* XXVI-4 (October 1896): 139-140.
27. *Wood and Iron* XXXVI-2 (August 1901): 58.
28. *The Humboldt Times*, August 4, 1883; Humboldt County Records, *Deeds*, Book 9, 760.
29. Howard Brett Melendy, "One Hundred Years of the Redwood Lumber Industry, 1850-1950" (Ph.D. dissertation, Stanford University, 1952), 86.
30. *Report of the Commissioner of the General Land Office, 1888*, 47.
31. S. A. D. Puter and Horace Stevens, *Looters of the Public Domain* (Portland, Oregon: The Portland Printing House, 1908), 18.
32. Ibid.
33. Ibid.
34. Humboldt County Clerk, *Articles of Incorporation*.
35. *The Humboldt Times*, April 19, 1888; Melendy, "One Hundred Years," 167-168.
36. *The Humboldt Times*, September 12, 1916; Robert Barnum, timber broker, private interview, Eureka, California, March 16, 1965.
37. Humboldt County Clerk, *Articles of Incorporation*.
38. *The Humboldt Times*, November 18, 1882.
39. *The Humboldt Times*, June 19, 1886 and March 22, 1888.
40. *Wood and Iron* XXIV-2 (August 1895): 60; XXIV-3 (September 1895): 68; XXIX-5 (May 1898): 178.
41. Claudia Wood, "The History of The Pacific Lumber Company" (term paper, Humboldt State College, 1956), 19.
42. *Wood and Iron* XXXVIII-4 (October 1902): 18.
43. *Wood and Iron* XXXIX-4 (April 1903): 18; XL-3 (September 1903): 17; XL-5 (November 1903): 23; XLII-3 (September 1904): 22.
44. Melendy, "One Hundred Years," 195.
45. Wood, "History of The Pacific Lumber Company," 22-31; Andrew Genzoli, "A Century in the Redwoods," *The Times-Standard*, May 27, 1969; *The Humboldt Times*, March 3, 1910 and March 5, 1910.
46. Melendy, "One Hundred Years," 215.
47. Stanwood A. Murphy, TPL, private interview, Scotia, California, May 5, 1972; *The Arcata-Union*, August 10, 1972.
48. *The Times-Standard*, February 27, 1970.
49. Ibid.; *The Arcata-Union*, March 1, 1978.

The Louisiana-Pacific plant on the peninsula opposite Eureka on Humboldt Bay. At the top of this 1976 picture the town of Samoa, the docks with oceangoing vessels, the dry kilns and lumber stacks, the planning mill, the mill and mill pond, the plywood plant, and the pulp mill at the lower portions of the photograph. — DAVID SWANLUND

Redwood
Capitalists and Companies:
The Second Wave

At the turn of the century, there was not enough money on the northern Pacific Coast to meet the capital needs of big business; but investors and bankers from other parts of the nation had already realized the opportunities waiting there, and more and more of them began to take a hand in the affairs of the Redwood Country. The Pacific Lumber Company's mortgage was typical of the times. In 1913 TPL put up sawmills, equipment, and 45,584 acres of timber as collateral for a loan that would help the company finance its program at Scotia. The Continental Trust & Savings Bank of Chicago and the Michigan Trust Company of Grand Rapids were the trustees.[1]

Andrew Benino Hammond, founder of the Hammond Lumber Company, began his meteoric career in Montana, when he contracted to furnish all the ties and trestle and tunnel timber for the Northern Pacific Railroad, which would connect that state with the East. Later, he began a successful lumbering operation on the Big Blackfoot River. After acquiring both the Missoula Water Works and the First National Bank of Missoula, he bought the Oregon Western & Pacific Railroad, and then built the Astoria & Columbia Railroad. Even before 1900, when he purchased the Vance Lumber Com-

pany, Hammond had secured valuable timberlands near Astoria and in the Cascade Mountains and was operating no fewer than three sawmills in Oregon.[2]

The Vance properties, which A. B. Hammond bought for $1 million, consisted of several thousand acres of redwood timber, a sawmill, a shingle and shake mill, the Eureka & Klamath River Railroad, a steam vessel and several lumber schooners, and 1,200 feet of waterfront. Hammond incorporated the Vance mill in New Jersey, and in September 1900, the Samoa facility, the largest redwood mill of that time, began producing lumber as the Vance Redwood Lumber Company.[3]

In May 1902, Hammond purchased 36,000 acres of timber on Redwood and Prairie creeks from the American Lumber Company. The tract contained over 2,500,000,000 board feet of standing redwoods, and the amount of money that changed hands made the purchase one of the largest single transactions in timber ever to be completed on the Pacific Coast.[4]

As president of the Vance Redwood Lumber Company from 1900 on, Hammond devoted every waking hour to developing and enlarging his new enterprise. Ahead of his time in many ways, he

Horse power transportation used to move lumber around the yard at the Vance Lumber Company (Hammond) on the peninsula in the early 1900's. — OTTO AND MARY EMILY DICK — G.J. SPEIER COLLECTION

Andrew B. Hammond

Hammond bought the Vance properties in 1900. The company survived as the Hammond Lumber Company until 1956 when it was sold to the Georgia-Pacific Corporation.

The Hammond Lumber Company office crew, plant bosses, and store personnel in 1932. Fred Cabral, sawmill boss (first row, second from far right) worked for the company for 50 years. — ARTHUR PETERSON COLLECTION

The factory workers of the Sash and Door Department of the Vance Lumber Company (Hammond) pose for the photographer in 1909. The late Andrew Kovacovich (far left, front row, kneeling) went to work for the company in 1909 and retired in 1965. — DONALD MAHLER COLLECTION

An overview of a portion of the Hammond Lumber Company town of Samoa on the peninsula about 1910. Sailing vessels and steam freighters are loading at the wharves on Humboldt Bay. The place was originally known as Brownsville. In 1889, a group of Eureka businessmen formed the Samoa Land & Improvement Company, so named because of the crisis in the Samoan Islands at the time and its similar appearance. The lumber town was called Samoa; and remains the center of lumber operations. — LOUISIANA-PACIFIC CORPORATION COLLECTION

(LEFT) A cook uses a gut hammer to call the woodsmen to dinner at a Hammond camp in the 1920's. — G.J. SPEIER COLLECTION (ABOVE) Three waitresses pose before serving the hungry loggers. — PAUL JADRO COLLECTION

The Little River Redwood Company mill at Crannell in the late 1920's. The company, which was incorporated in 1892, merged with the Lagoon Lumber Company which added extensive timberlands in the Big Lagoon area. The new Little River Redwood Company was incorporated in 1923. From 1924 to 1931, the company lost $4 million. Faced with bankruptcy, the company merged with the Hammond Lumber Company whose interests controlled the two firms. The Hammond and Little River Redwood Company Ltd. was incorporated in 1931, but two years later the name was again changed to Hammond Redwood Company. The Crannell mill closed down on July 21, 1931. — PAUL JADRO COLLECTION

Leonard C. Hammond

After his father Andrew B. Hammond died in 1934, Leonard C. Hammond became president of the Hammond Lumber Company, the Hammond Shipping Company, and the Redwood Export Company of San Francisco. Leonard died in 1945. — OREGON HISTORICAL SOCIETY COLLECTION

was among the first employers in the region to hire minorities. In 1905 his company expanded its California holdings by purchasing property in Los Angeles, and in 1912 it was re-named the Hammond Lumber Company.[5]

The Little River Redwood Company (Humboldt County), which had begun operations in 1908, merged with Hammond Lumber in 1931, and the directors of the latter controlled both firms. For a time, the company was known as the Hammond & Little River Company, but in 1933 it became the Hammond Redwood Company. In 1937 Hammond Redwood bought the Humboldt Redwood Company's mill in Eureka, and that facility became its Plant Number Two.[6]

A. B. Hammond was eighty-five years old when he died, in 1934. His son, Leonard C. Hammond, who succeeded him as president of the Hammond Redwood Company, died in 1945, at the age of sixty-one.[7]

In 1931, redwood lumber executives pose in front of a Southern Pacific freight train consisting of 50 cars of redwood. The train was intended to promote redwood to the eastern lumber market. A.B. Hammond is the man with the beard, and Governor James Rolph, Jr. is dressed as the engineer. — GEORGE KNAB COLLECTION

Alfred Bell, Jr.

Alfred Bell, Jr., Leonard Hammond's stepson, arrived in Samoa in 1921. His first job in 1924 was cleaning up behind the horses. After graduating from Harvard, he went to work for the Hammond Lumber Company, advancing to Sales Manager. In 1940 he was transferred to San Francisco as Eastern Sales Manager. In 1947 Alfred bought 51 percent in the Hobbs, Wall & Company in San Francisco, and with this company he became President and Chairman until his retirement in 1975.

In 1937 Hammond Lumber Company purchased the Bayside mill in Eureka from the Humboldt Redwood Company to handle the logs from their Van Duzen operations. This mill became their Plant Two. In 1962 the Georgia-Pacific Corporation closed the mill down and sold the property. — JOHN AND MARY JANE PHEGLEY COLLECTION

The Northern Redwood Company at Korbel closed its doors in 1933, and its effort to resume functioning later that year was unsuccessful. In 1940, however, it was able to borrow $1 million from the Reconstruction Finance Corporation, and Fentress Hill, a San Francisco financier, reorganized the company. Northern Redwood was back in operation by 1942.[8]

****

The Arcata Redwood Company (ARCO) was founded in 1939 with a capital investment of $50,000, augmented by an additional $24,250 the following year. Its president until he retired in 1967 was Howard Libbey, who had formerly been assistant general manager of the Little River Redwood Company (Humboldt County) and general manager of Hobbs, Wall & Company of Crescent City (Del Norte County) and San Francisco.[9]

In 1946 ARCO issued 116,690 shares of stock to its sole stockholder, the Hill-Davis Company, Ltd., a Michigan partnership. Years earlier, in 1909, Hill-Davis had secured timber on the north bank of Redwood Creek, and by 1931 it controlled close to 40,000 acres of timber in Humboldt County. In 1958 Hill-Davis exchanged timberlands that bordered Georgia-Pacific (G-P) lands for stock in the Arcata Redwood Company, thus bringing about a merger of ARCO's mill operation with its own timber holdings. The merger set the stage for expansion that included development of a sizeable logging operation and construction of two sawmills near Orick. The company also built a manufacturing plant on Humboldt Bay, between Eureka and Arcata.[10]

In 1964 Arcata National Company, a wholly-owned subsidiary of ARCO, bought the J. W. Clement Company of Buffalo, one of the nation's biggest printing firms, for about $13.5 million. J. W. Clement, operating through several subsidiaries of its own, including Pacific Press of Los Angeles and Phillips & Van Orden of San Francisco, prints *Reader's Digest* and most West Coast editions of major national magazines. In 1966 the Arcata National Company, under a new Board of Directors and with headquarters in Menlo Park, became the parent company and ARCO became its subsidiary.[11]

George E. Knab

In 1919 George E. Knab started to work for the Babcock Lumber Company in Pittsburgh, Pennsylvania. Three years later he hired on with the California & Oregon Lumber Company in Brookings, Oregon. After the dissolution of this company, he joined The Little River Redwood Company of Crannell in 1926. This company merged with the Hammond Lumber Company, and he moved to the Samoa office. In 1934 Len Hammond sent George to the Chicago office as Sales Manager where he remained until he joined Howard Libbey, President of the Arcata Redwood Company, in 1947. Until his retirement in 1964, George was the Eastern Sales Manager for ARCO.

The Arcata Redwood Company (ARCO) mill A, just north of Orick. — DAVID SWANLUND

Arcata Redwood Company's plant at Brainard on Humboldt Bay between Eureka and Arcata. Located here are the dry yard, planing mill, dry kilns, shipping yards, and offices. The ARCO plant began operating here in 1952. — DAVID SWANLUND

Howard A. Libbey

Howard A. Libbey, a native Eurekan, went to work for The Little River Redwood Company in 1916. In 1934 he joined Hobbs, Wall & Company and was appointed Vice President and General Manager at Crescent City. In 1939 he formed the Arcata Redwood Company and remained president of the company until his retirement in 1967. — ARCATA REDWOOD COMPANY COLLECTION

Burnette Henry

Burnette Henry, President of ARCO, graduated from the Missouri School of Mines. After World War II, he held positions with the U.S. Gypsum Company in Houston, Acme Brick Company in Fort Worth, and the Republic Housing Corporation in Dallas. In 1974 Burnette joined Arcata National as Manager of Engineering Operations and became Vice President of Manufacturing for the Book Group. In 1975 he was promoted to Vice President of Operations of the Arcata Book Group in Tennessee. In 1977 he was elected president of Arcata Redwood and Vice President of Arcata National by the Board of Directors in Menlo Park. — ARCATA REDWOOD COMPANY COLLECTION

****

After World War II, several newcomers joined the established companies in Humboldt County. The biggest of these were Simpson Timber Company; Eureka Plywood Company, valued at $2 million; Arcata Plywood; Mutual Plywood, valued at $1.75 million; and Roddiscraft, Inc. The Weyerhauser Timber Company of Tacoma, Washington, took over Roddiscraft in 1960. In December 1965, however, Weyerhauser sold off its 36,000 acres of timberland in Humboldt, Del Norte, and Mendocino Counties, having found that to continue the Arcata operation would be impossible. The Weyerhauser particleboard plant, plywood mill, and logging equipment at Arcata were sold to Humboldt Flakeboard, a new California corporation.[12]

The picturesque setting of the Simpson Timber Company plant at Korbel on the North Fork of the Mad River. — DAVID SWANLUND

The Simpson Logging Company of Washington (now the Simpson Timber Company) entered the Redwood Country in 1948 with the purchase of a sawmill at Klamath and 23,000 acres of timber valued at $9 million. On April 1, 1956, it purchased the Northern Redwood Company and built a new mill at Klamath, as well as a huge re-manufacturing plant, which included a planing mill and drying yards, just west of Arcata near the ocean.[13] The following summer, it acquired all the remaining timber holdings of the Sage Land and Improvement Company.[14] Later, in 1965, Simpson Timber purchased the plywood plant at Arcata and the timber holdings in Humboldt, Del Norte, and Mendocino Counties that had formerly been owned by the Weyerhauser Company. The transaction involved 35,000 acres of cutover and new growth land and cutting rights to more than 100 million board feet of old growth, mainly Douglas fir. The deal increased the amount of Redwood Country timberland owned by Simpson Timber to 245,000 acres.[15]

In October 1967, Simpson Timber acquired the Mutual Plywood plant at Fairhaven from United States Plywood-Champion Papers, Inc., plus 3,400 acres of second growth timberland in the Bald Hills area, east of Orick, and first-right options to purchase 250 million board feet of Douglas fir in the eastern portion of Humboldt County.[16] In 1969 the company completed a $5 million modernization program at Korbel, which included new equipment, a new building, and a new resaw facility. Today,

its former log pond has given way to a dry pond operation.[17]

The Simpson Timber Company, in 1978, deplored the decision of Congress to expand Redwood National Park. Nevertheless, its president, Gilbert Oswald, announced a "long-term commitment to continue operations in California despite land grabs and excessive regulation by the State of California." He went on to say that, as an example of that commitment, Simpson Timber was initiating a $700,000 program of environmental improvements at its Mad River plywood plant and a $1.2 million investment in reforestation and intensive forest management that would include outplanting 4.5 million seedlings from its Korbel nursery in both Humboldt and Del Norte counties.[18]

Simpson is currently in the midst of a $57 million capital improvement program for its California operations, to be completed by the end of 1981, of which $33 million is to upgrade and modernize the mills at Klamath and Korbel and the plywood plant at Fairhaven. The remainder is for roads and new logging equipment, as the company moves further into its young growth management program.[19]

Henry Trobitz

Henry "Hank" Trobitz, a graduate of the University of California, joined the Simpson Company at Shelton, Washington in 1948 where he was involved in photogrammetry. In 1950 Hank came down to Klamath where he was Timberlands Manager. In 1956 he transferred to Arcata where he became Timberlands Manager in charge of all forestry and logging. In 1972 he became California Resources Manager. Hank retired in 1981. — SIMPSON TIMBER COMPANY COLLECTION

Pent up for more than a decade because of the Depression and World War II, the nation's demand for lumber rose sky high in 1945, to its highest level in forty years. To meet these soaring requirements, lumbermen had to go searching for undeveloped timber areas, since the amount of easily available old growth timber in the Pacific Northwest was being rapidly depleted. Many of them headed for Humboldt County, drawn by its untapped stands of Douglas fir, and the result was spectacular. By 1953 the county's lumber production was four and one-half times greater than it had been in 1940, and the number of its sawmills had increased eightfold. Where before the war Arcata had been subjected to the disparaging "Arcata potater, two stores, and one theater," billboards now proclaimed it the "lumber center of the world." Along with redwood, Douglas fir is now the raw material of a major Humboldt industry. From it comes a different type of lumber product, directed to different markets, giving rise to different manufacturing techniques, and calling for a different pattern of industrial organization than the monarchs of the forest.

****

On May 24, 1956, the local newspapers in the Humboldt region carried the big news that the Hammond Lumber Company had been sold to the Georgia-Pacific Corporation of Atlanta, Georgia for $80 million. The Hammond Lumber Company was one of the two largest redwood mills in existence at the time, and the sale was the biggest financial deal in the history of Humboldt County.[20] For the previous year, Hammond plants' production in board feet was as follows: Plant One (Samoa), 36,030,000; Plant Two (Eureka), 41,102,000; Plant Three (Big Lagoon), 5,520,000; and Plant Four (Orick), 1,607,300. The total equaled 84,260,300 board feet of lumber. At the time of the sale, Hammond Lumber Company's timber holdings broke down as follows: Redwood Creek Block, 1,005,253,245 feet; Big Lagoon Block, 1,192,265,732 feet; Van Duzen Block, 597,306,291 feet; Eel River Block, 237,080,831 feet; Geneva Block, 228,280,000 feet. The total was 3,260,186,099 feet.[21]

The Georgia-Pacific Corporation (G-P) took over the assets of Hammond Lumber on October 22, 1956, and the redwood operations became the Hammond-California Redwood Company, a wholly-owned subsidiary of G-P. Georgia-Pacific was not interested in the Hammond Lumber Company's thirty-three retail lumber yards and building supply stores, and these were later either sold or liquidated. Soon after the take-over, G-P closed Plant Four at Orick. The mill was torn down, and

Harry Merlo

Harry Merlo, Chairman and President of the Louisiana-Pacific Corporation, was named a recipient of the Horatio Alger award in 1980. Under his leadership the company has grown in sales from an initial $272 million to more than $1.3 billion in 1979. — LOUISIANA-PACIFIC CORPORATION COLLECTION

Arthur Peterson

Arthur "Bud" Peterson, who came from Kerkhoven, Minnesota, started to work for the Hammond Lumber Company in 1926 and worked his way up to production superintendent for the Hammond Lumber Company and the Georgia-Pacific Corporation. Bud retired in 1971. — LOUISIANA-PACIFIC CORPORATION COLLECTION

the Arcata Redwood Company bought the land.[22]

To help finance its acquisitions, G-P sold some of its timberlands. The major portion of the approximately 237,080,000 feet of timber in the Eel River Block was purchased by the Willits Redwood Products Company, which logged the timber and then shipped the logs to Willits (Mendocino County) by rail. Other timber and timberlands in the Van Duzen Block were bought by various companies, including Tidewater Lumber, Inc., Van de Nor Lumber Company, and the Rockport Redwood Company of Mendocino County. Rockport Redwood bought only 961 acres. From July 1956 to November 30, 1957, G-P divested itself of close to 1,500,000,000 feet of timber for about $45 million.[23]

In 1958 G-P began building a plywood mill at Samoa, in order to make use of the Douglas fir logs on its property. The mill opened in 1959, and G-P also built a stud mill to produce studs from plywood cores. By 1960 the corporation closed Plant Two in Eureka, and by the spring of 1962, all the lumber in the modern remanufacturing plant and the new

The Hammond Lumber Company plant and the town of Samoa on the peninsula in 1947. — MERLE SHUSTER AND JOHN PHEGLEY COLLECTION

The Humboldt Peninsula located across the bay from the City of Eureka. At the top of this scene the town of Samoa, the Lousiana-Pacific Lumber plant, the L-P plywood mill, the L-P pulp mill, the Crown-Simpson pulp mill, and at the lower portion the Simpson Timber Company's plywood plant at Fairhaven. — DAVID SWANLUND

dry yard had been sold or trucked to Samoa. The property was purchased by the Park Loading Company of Portland. In 1963 G-P began construction of a 55-ton per day bleached kraft pulp mill and a new automated redwood lumber mill. The latter began production August 14, 1965. The old sawmill built by the Vance family in 1894 and remodeled by Hammond Lumber in 1924 and 1925 was shut down December 18, 1964.[24]

The new automated redwood mill used 345,000 feet of logs each day to produce 400,000 board feet of lumber and required approximately 100 truckloads of logs daily. The mill boasted a hydraulic barker and a head rig with a ten-foot diameter bandmill. The plant had two edgers, a re-edger, a cant gang saw capable of sawing 100,000 to 130,000 board feet each shift, two resaws, and three gang trim saws. On an average day, eighty-two men were employed there, fourteen of them on maintenance. The last of the old growth redwood logs made its way through this mill on Friday, February 1, 1980. A new small-log mill, built to replace the old growth mill and capable of turning out 90 million board feet a year, began operating less than two weeks later.[25]

On January 11, 1967, G-P purchased the historic 8,000-acre McKay tract and its stand of young growth timber just southeast of Eureka for "several millions of dollars" from the Pacific Conservation Company, whose principals were G. Kelton Steele of Eureka and Herbert Malarky of Portland. Logged by bull teams during the 1890's, the tract is one of the finest stands of second growth in the redwood region.[26]

When the Rockport Redwood Company on the Mendocino coast closed down its sawmill in 1957, its Cloverdale plant assumed the name Rounds & Kilpatrick Lumber Company. Harry A. Merlo, a forester and graduate of the University of California at Berkeley, was its vice-president and general manager. In 1967 G-P purchased Rounds & Kilpatrick, "mainly to get Harry."[27] Merlo moved to Samoa and became vice-president and general manager in charge of G-P's timber, plywood, and lumber operations there. In August 1968, to enhance G-P's position in the Redwood Country, he purchased F. M. Crawford Lumber, Inc., which included sawmills at Alderpoint and Dinsmore in Humboldt County and at Ukiah, Willits, Potter Valley, and Covelo in Mendocino County, as well as a remanufacturing plant at Calpella in Mendocino County, for $9,360,000.[28]

By September 1968, in addition to the Rounds Industries, which included the Cloverdale plant, the 36,841-acre Ralph M. Rounds Tree Farm near Rockport (Mendocino County), and 2,000 acres south of Rockport, G-P had also acquired the Aborigine Lumber Company and a stud mill at Fort Bragg in Mendocino County.[29]

Alfred H. Merrill

Alfred H. Merrill graduated from the University of California with a degree in Forestry. After working for two different lumber companies, he joined the Hammond Lumber Company in 1947. When Georgia-Pacific bought the Hammond company in 1956, Al became Chief Forester, and later when Louisiana-Pacific took over in 1973, Al continued in the same position. On April 24, 1979, Harry Merlo promoted Al to the position of California Manager of Forestry Affairs for L-P.

Charles C. Barnum

Charles "Charlie" Barnum was the most successful timber broker in the Redwood Country. In addition to his own business, Charlie, in partnership with G. Kelton Steele in the Humboldt Lands Company, sold the following: the Redwood Creek Ranch with good timber; the McKay tract in the Ryan Slough basin, and the Merriman tract on the west slope of Redwood Creek. Since Charlie's death in 1953, his son, C. Robert Barnum has built the Barnum Timber Company into a $20 million concern which controls 53,000 acres of timber in Humboldt County.

The Samoa Division of G-P at this time consisted of the Samoa sawmill, plywood mill, stud mill, and pulp mill; the Big Lagoon sawmill; the Carlotta sawmill; Humboldt County timberlands at Big Lagoon and on the Van Duzen River; cutover land on the Eel River; the six basically fir and pine sawmills of F. M. Crawford; the Aborigine mill in Fort Bragg; the Rockport lands; the Cloverdale plant; and the Feather Falls sawmill in Butte County, purchased in 1955.[30]

In 1969 G-P acquired Tidewater Lumber, Inc., in Eureka, which had a sawmill and docks on Humboldt Bay. It closed down the mill in 1971, but kept the docks. In May 1970, it bought the Brightwood Lumber Company sawmill near Arcata and the All-Brite Lumber Company, a remanufacturing plant also near Arcata.[31]

Harry Merlo was promoted in 1971 to the Portland headquarters of G-P as executive vice-president in charge of lumber and plywood on the West Coast, as well as of G-P's Alaskan operations.[32]

At the time it acquired the Hammond Lumber Company, in 1956, the Georgia-Pacific Corporation had only a few sawmills and plywood mills in the South and in the Pacific Northwest. Through mergers, acquisitions, and internal expansion, G-P developed into a giant among forest products

Philip E. Nell

Philip E. Nell went to work with the Red River Lumber Company at Canby, California, and later became co-owner of the Indian Creek Lumber Company in La Grande, Oregon. From there he came to Healdsburg, California where he became Manager of Sonoma Wood Products. In 1971 the Georgia-Pacific Corporation at Samoa hired Phil to be General Manager in charge of building products, and in 1973 he became General Manager for the Louisiana-Pacific Corporation at Samoa. Phil retired in 1981.

The last of the old-growth redwood logs made its way through Louisiana-Pacific's Samoa sawmill on February 1, 1980, marking the end of such operations for the company at Samoa. L-P lost much of its old-growth timber in the expansion of the Redwood National Park.
— LOUISIANA-PACIFIC CORPORATION COLLECTION

companies — such a giant that in June 1971 the Federal Trade Commission issued a complaint against the corporation's purchase of sixteen firms and 630,000 acres of southern pine trees and tried to prevent G-P from making any more acquisitions in the forest products industry for the next ten years.[33]

On July 24, 1972, however, G-P and the Federal Trade Commission reached an agreement. The Louisiana-Pacific Corporation (L-P), a new Delaware corporation, was put together from pieces of G-P. Its base was in California, in the Ukiah and Samoa divisions, and its headquarters was in Portland. The Ukiah Division included the sawmills at Ukiah, Willits, Alderpoint, Dinsmore, Potter Valley and Covelo, and the remanufacturing plant at Calpella, all purchased from F. M. Crawford Lumber in 1968, plus sawmills at Fort Bragg and Oroville (Butte County). It also had approximately 157,000 acres of timberland in Mendocino County, primarily Douglas fir. The Samoa Division included sawmills at Samoa, Carlotta, and Big Lagoon; remanufacturing plants at Eureka, Cloverdale, and Healdsburg (Sonoma County); the pulp and plywood mills at Samoa; and 125,000 acres of timberland.[34]

Soon after it had spun off the Samoa Division, G-P proceeded to purchase the former Union Lumber Company redwood mill in Fort Bragg from the Boise Cascade Corporation. Prohibited by the Federal Trade Commission settlement from acquiring any plywood plants, it sold the Fort Bragg plywood mill to L-P, along with approximately 45,000 acres of Mendocino timberland.[35]

The Louisiana-Pacific Corporation was created and spun off from the Georgia-Pacific Corporation in January 1973. William H. Hunt, president of G-P from 1970 to 1972, became the first Chairman of the Board of Louisiana-Pacific. Harry Merlo, however, who became L-P's first president at the age of fifty-two, was the driving force. In 1973 alone he made fourteen acquisitions, and L-P, having already outdistanced Boise Cascade, is racing to catch up with the Weyerhauser Company and win for itself the title of "largest timber company in the United States."[36]

****

Signs of the future in the lumber industry today are the bleached kraft pulp mills on the Humboldt Bay Peninsula, opposite the city of Eureka. These enormous plants and their towering piles of chips dominate the landscape and are visible for miles around. The Georgia-Pacific Corporation's mill, the first to be designed and constructed specifically for utilization of redwood chips as a source of raw material for bleached kraft pulp, began operating in 1965 and was taken over in 1973 by the newer Louisiana-Pacific Corporation.[37]

The Crown Simpson Pulp Mill at Fairhaven, just south of the Louisiana-Pacific plant, was built in 1966. It represents a partnership on the part of two of the Northwest's major forest products companies — Crown Zellerbach Corporation, second largest paper manufacturer in the world, and the Seattle-based Simpson Timber Company. Constructed at an approximate cost of $50 million, the Fairhaven mill can turn out 500 tons a day of bleached kraft pulp. Simpson Lumber, which supplies the mill with chips, and Crown Zellerbach, whose personnel manage and operate it, each receive half of its production for separate sale to the world market.[38]

The pulp manufactured by both the Louisiana-Pacific and the Crown Simpson pulp mills is produced entirely from redwood and Douglas fir chips generated by lumber and plywood mill operations in northern California that would once have been burned as unusable material, but are now transported to the pulp mills by lumber companies' trucks or independent trucking contractors. Thus the pulp industry has provided Humboldt and Del Norte county lumber and plywood suppliers with a major new source of revenue. Approximately 120 trucks per day arrive at the Crown Simpson plant chip dump alone, and that company buys about $4.8 million worth of wood chips annually — clear confirmation of the tremendous economic significance of this new development in the Redwood Country.[39]

Chapter Notes

1. Howard Brett Melendy, "One Hundred Years of the Redwood Lumber Industry, 1850-1950" (Ph.D. dissertation, Stanford University, 1952), 202.
2. *Hammond Redwood Log* 3 (March 1950).
3. Ibid.
4. *The Humboldt Times*, May 17, 1902.
5. *Wood and Iron* XLIII-1 (January 1905): 10; *Hammond Redwood Log* 3 (March 1950).
6. *The Humboldt Times*, September 30, 1936; *Hammond Redwood Log* 3 (March 1950).
7. *The Humboldt Times*, January 6, 1934 and December 21, 1945.
8. Lynwood Carranco and Mrs. Eugene Fountain, "California's First Railroad: The Union Plank Walk, Rail Track, and Wharf Company Railroad," *Journal of the West* III (April 1964): 252.
9. William Bronson, "Behind the Redwood Curtain," *Cry California* I-4 (Fall 1966): 10; Howard A. Libbey, retired president of ARCO, private interview, Eureka, California, May 22, 1972.
10. Ibid.
11. Howard A. Libbey, private interview, May 22, 1972.
12. *The Humboldt Times*, December 7, 1965.
13. *The Humboldt Standard*, March 30, 1949; James Hartley, Simpson Timber Company, private interview, Arcata, California, September 26, 1970; *The Times-Standard*, March 25, 1977.
14. Robert Barnum, timber broker, private interview, Eureka, California, March 16, 1965; Henry K. Trobitz, private interview, Arcata, California, December 6, 1980.
15. *The Humboldt Times*, December 8, 1965.
16. *The Times-Standard*, October 10, 1967.
17. *The Arcata Union*, December 4, 1969.
18. *The Arcata Union*, March 1, 1978.
19. Henry K. Trobitz, private interview, Arcata, California, December 6, 1980.
20. *The Humboldt Times*, May 24, 1956.
21. "California Lands," *General Information Book*, 2-2, 3-2, 4-2, 5-2, 7-1; Lowell S. Mengel II, "A History of the Samoa Division of Louisiana-Pacific Corporation and Its Predecessors, 1853-1973" (Master's thesis, Humboldt State University, 1974), 175-181.
22. Mengel, "A History of the Samoa Division," 175-181.
23. Ibid., 10-12.
24. Arthur Peterson, Georgia-Pacific Corporation, private interview, Samoa, Calfiornia, June 4, 1965.
25. *Welcome to the Sawmill*, published by the Georgia-Pacific Corporation, 1-3.
26. *The Times-Standard*, January 12, 1967.
27. *Business Week*, October 13, 1973, 122.
28. *The Times-Standard*, April 27, 1968.
29. *Second Growth* (Samoa Division newspaper), September 1968, 1.
30. Ibid.
31. *The Times-Standard*, February 24, 1971.
32. *The Times-Standard*, February 11, 1971.
33. Mengel, "A History of the Samoa Division," 199-200.
34. Ibid., 201-202.
35. Lois Bishop, Louisiana-Pacific Public Relations, private interview, Samoa, California, June 5, 1973.
36. *Business Week*, October 13, 1973, 122.
37. Arthur Peterson, Georgia-Pacific Corporation, private interview, Samoa, California, May 5, 1969.
38. *Welcome to Fairhaven* (Fairhaven, California: Crown Simpson Pulp Company).
39. *Background on Fairhaven Pulp Mill* (Fairhaven, California: Crown Simpson Pulp Company), 2-3.

The Mendocino Lumber Company's second mill on the flat near the mouth of Big River in the late 1880's. The original Big River mill was dismantled in 1857. — NANNIE M. ESCOLA COLLECTION

Redwood Mills in Mendocino County

Of the three counties in the Redwood Country, Mendocino is the largest. Redwood mills there were built along its rugged coast from Usal on the north to the Gualala River in the south. Since rivers were the only means of entry to the timber belt, where the trees as a rule grew on steep slopes, they became a determining geographic factor in the development of the Mendocino redwood industry.

****

The earliest sawmill on the Mendocino coast was built in 1852 at Big River by E. C. Williams, Jerome Ford, Henry Meiggs, and Captain David Lansing, all lumbermen from New England who had come to California in search of gold. They called their new enterprise the California Lumber Company, and in 1853 they constructed a second mill on the flat near the mouth of the Big River, in order to increase production. This mill was equipped with two single circular saws, one muley saw, and one sash saw, and it could cut 40,000 board feet a day.[1]

In 1854 Henry Meiggs, who directed California Lumber Company activities from San Francisco and held a substantial interest in the firm, fell into financial straits and left the country. Meiggs had borrowed heavily from the private banking firm of Godeffroy & Sillem, pledging his California Lumber Company stock as collateral. On the basis of that pledge, the banking firm, after Meiggs' hasty departure from the United States, exercised its right to a three-fifths interest in the fledgling lumber firm. However, Godeffroy & Sillem also raised additional capital to help Williams and Ford, and these two hard-working men eventually paid off the indebtedness.[2]

California Lumber dismantled its original Big River mill in 1857. The second mill, at the mouth of the Big River, burned in 1863, but the company built a new one the following year.[3] By 1873 the firm had reincorporated under the name of the Mendocino Lumber Company. Valued at $60,000, it was the most important mill in the county.[4]

In 1880 it was still leading the field. Mendocino County assessment records for that year valued it at $166,533, a higher figure than was listed for any other Mendocino mill. J. B. Ford, a native of Vermont, served as resident manager of California Lumber until 1872, shortly before its reincorporation, when he moved to Oakland. Ford died in

The Mendocino Lumber Company's second mill at the mouth of Big River before it burned down in 1863.
— NANNIE M. ESCOLA COLLECTION

1889. In 1896, when E. C. Williams, the only one of the company's founders still living, was president of the firm, Mendocino Lumber's old-fashioned mill was still turning out a good grade of lumber.[5]

The Ford family retained its stock in the company until 1902, when Captain A. M. Simpson, of Simpson Lumber and the Crescent City Mill & Transportation Company (Del Norte County) bought out their interest.[6] In November 1904, the old circular saws were removed from the mill, and

Charles R. Johnson

Charles Russell Johnson, the founder of the Union Lumber Company at Fort Bragg. — EDWARD FREITAS COLLECTION

a bandmill was installed, and by 1905 Mendocino Lumber was cutting 100,000 board feet daily.[7]

C. R. Johnson of the Union Lumber Company and C. J. Wood of the Caspar Lumber Company took over Mendocino Lumber in 1906, and the pioneering firm's existence as an independent entity came to an end. However, although Union Lumber now controlled fifty-two percent of the stock, no immediate organizational or operational changes took place.[8] Not until 1912 did the new owners dissolve the organizational structure that had resulted from the pioneer Mendocino Lumber Company's reincorporation in 1873, and even then they retained the firm's name. E. C. Williams died the following year, at the age of ninety-three.[9]

Mendocino Lumber, the oldest firm in the redwood industry, was forced to close in 1931 by the diminishing lumber market of the Depression years. Its mill operated for the last time in 1938, when a raft of fir logs en route from the Columbia River to San Diego broke up off the Mendocino coast. About half the raft was salvaged and hauled to the Big River mill, which turned it into 3,500,000 board feet of lumber. Shortly thereafter, the mill's equipment was dismantled and taken away.[10]

****

Charles Russell Johnson, widely referred to as C. R. Johnson, was twenty-two years old in 1881, when he decided to give up working for his father, a Wisconsin lumberman, and head for California. He arrived at Newport in Mendocino County the following year and purchased an interest in the Hunter & Stewart mill. In 1883 the company was re-named Stewart, Hunter & Johnson, and its new partner introduced the first successful night shift

An early view of the Union Lumber Company mill at Fort Bragg. C.R. Johnson built this mill in 1884, and the company lasted until 1969. Note the log train dumping logs into the millpond in the center of this scene. — NANNIE M. ESCOLA COLLECTION

operation in the county.[11]

The Newport mill was too small for the company's plans and had a poor loading point, so in 1884 C. R. Johnson chose the site of the former army post at Fort Bragg as a good location for a new mill, and the owners of Stewart, Hunter & Johnson formed the Fort Bragg Redwood Company, purchasing a large tract of timber from Macpherson & Wetherbee, who had earlier acquired vast holdings in the vicinity of Fort Bragg. The transaction included all of the present Union Lumber Company plant site, most of the land now comprising the city of Fort Bragg, and certain timberlands on Pudding Creek and the Noyo River. With financial support from his father and friends, Johnson's company was able to begin construction of a new mill that year, and the young man from Wisconsin, at the age of twenty-five, launched the most extensive project the Mendocino coast had ever seen. His lumber company would evolve into one of the three largest redwood operations in the world, and he would found and build the biggest coastal community between San Francisco and Eureka.[12]

C. R. Johnson started production at the new Fort Bragg mill in November 1885. The installation of a band saw paid off, and the Fort Bragg Redwood Company began to expand. In 1887 its directors were C. R. Johnson of Fort Bragg; O. R. Johnson of Racine, Wisconsin; and Governor Alger and Senator Stockbridge of Michigan. That same year, Fort Bragg Redwood secured controlling interest in the White & Plummer Company, which owned an old mill at Noyo, and formed the Noyo Lumber Company.[13]

When Fort Bragg Redwood's timberlands along Pudding Creek were logged out, the Noyo River area was opened to logging. The timber holdings of W. P. Plummer and C. L. White on the Noyo were merged with those of Fort Bragg Redwood in a transaction, completed August 17, 1891, that gave birth to the Union Lumber Company.[14]

In 1896 the Union Lumber Company cut one-third of Mendocino County's lumber output. Its

The White & Plummer Company Noyo mill (just south of Fort Bragg) in the 1860's. This company merged with Fort Bragg Redwood in 1891 to become the Union Lumber Company.
— CALIFORNIA REDWOOD ASSOCIATION COLLECTION

domestic shipments were loaded at Fort Bragg, and its foreign orders by wire chute at Noyo Harbor, just to the south. In San Francisco the company owned a planing mill that finished 50,000 board feet per day, and it had also acquired 60,000 acres of timber on the Noyo and Ten Mile rivers. In 1903 C. R. Johnson turned the mill at Pudding Creek, which had been idle for five years, into the Glen Blair Redwood Company, retaining a one-fourth interest in that firm.[15]

From its inception, the Union Lumber Company followed a policy of steadily increasing its production and expanding its facilities. By 1913, in addition to 65,000 acres of redwood timber near Fort Bragg, Union Lumber owned the California Western Railroad & Navigation Company, the National Steamship Company, the total stock of the Fort Bragg Electric Company, the West Coast Redwood Company, the Little Valley Lumber Company, fifty-two percent of the Mendocino Redwood Company, fifty percent of the Glen Blair Redwood Company, and eighteen percent of the Redwood Manufacturing Company.[16] Ten years later, in 1923, out of a combined total of 171 million board feet of redwood lumber cut by Mendocino County mills, Union Lumber and its affiliated mills at Glen Blair and Mendocino City cut 130 million board feet.[17]

The Great Depression caused most Mendocino mills to close down during the early 1930's. C. R. Johnson, however, managed to keep Union Lumber operating. The company weathered the bad times and continued along its progressive and prosperous road.[18] In 1939, when he was eighty years old, C. R. Johnson resigned as president of Union Lumber. The Board of Directors elected him chairman and elected his son, Otis R. Johnson, to succeed him in the presidency. Until his death in February of the following year, the redoubtable C. R. continued to be active in company affairs.[19] The Union Lumber Company continued to grow by purchasing the Usal tract of more than 12,500 acres from the Simpson Redwood Company in 1936, some 35,000 acres of Mendocino timberlands from the Pacific Coast Company in 1960, and the Seaside Lumber Company of Willits in April 1967.

Otis R. Johnson died July 1, 1957, and his son C. Russell Johnson became president of Union Lumber. Local control of the company, however, came to an end on January 15, 1969, when Union Lumber merged with the Boise Cascade Corporation of Boise, Idaho, becoming one of thirty-three companies that had merged with it since 1957. In 1973 the Georgia-Pacific Corporation purchased the former Union Lumber redwood mill in Fort Bragg from Boise Cascade.[20]

****

In 1861 William H. Kelley, acting for Kelley & Randall, purchased 10,340 acres of timberland along Caspar Creek in Mendocino County and then constructed a sawmill at the mouth of the stream. The mill was equipped with a sash saw and a single circular "pony saw," and had a daily cutting capacity of 15,000 board feet of lumber.[21] In 1864 Jacob Green Jackson, a Vermonter who had reached California in 1851, purchased the

The Caspar mill on the Caspar River at low tide in January 1869. — NANNIE M. ESCOLA COLLECTION

Kelley & Randall mill at Caspar Creek and renamed it the Jacob Green Mill. By adding a double-circular saw, he increased its daily output to 25,000 board feet. Jackson confined his company's early logging to the Caspar Creek basin and loaded finished lumber onto sailing schooners by means of barges. In 1874 Jackson built a railroad from his mill pond to Jug Handle Creek, in order to reach timber about a mile and a half from the mill. In March 1877, he bought timberland on Hare Creek from A. W. Macpherson. In 1878 he started construction of a new logging camp on Jug Handle Creek and extended his railroad, by then called the Caspar Creek Railroad, to a total distance of two miles, in order to reach the new camp.[22]

Within two years, Jackson was employing 166 men at Caspar, and the mill's daily capacity had increased to 45,000 board feet. In addition, the Jacob Green enterprise had a shingle mill and four logging camps. The railroad included one locomotive, ten cars, and 3.5 miles of track, and was worth $45,000. At this time, Jackson also owned four schooners. Shipments were loaded onto them by means of a chute connected to the mill by a tramway.[23]

In 1880 Jackson incorporated his holdings as the Caspar Lumber Company, which issued 4,000 shares of stock on a capital investment of $400,000. In July of the following year, he sold the Caspar Creek mill, timber, and railroad to this new corporation for $265,938.11 and 399 shares of stock.[24]

When fire destroyed the old mill at the mouth of Caspar Creek in 1889, Caspar Lumber constructed another, replacing the former circular saws with two band saws, thus raising daily cutting capacity to 90,000 board feet.[25] In 1893 the company bought a tract of timberland on the south fork of the Noyo River from the Union Lumber Company for $20,000. That same year, Casimer Jackson Wood became a director, and in 1899 Mrs. Abbie E. Krebs was also made a director.[26]

Jacob Green Jackson died April 17, 1901, at the age of eighty-four.[27] Two weeks later, on May 1, 1901, Henry Krebs was elected president of Caspar Lumber, Mrs. Abbie E. Krebs became vice-president, and Mrs. Elvenia D. Jackson replaced her as a director. In succeeding years, the heirs of Jacob Green Jackson continued to run the company. Dissimilarities in their surnames warrant the following clarification: Jacob Green Jackson and his wife, Elvenia D. Jackson, had one son, Charles Green Jackson, and one daughter, Abbie E. Jackson. Charles went into business on his own and was not associated with Caspar Lumber. Abbie Jackson married four times. Clarence E. DeCamp was her son by her first husband. Casimer Jackson Wood was a son of her second marriage. Captain Henry Krebs was her third husband, and F. A. Wilkins was her fourth husband.[28]

In 1901 also, the company purchased two separate parcels of timberland, totaling 5,200 acres, from William Heezer and Samuel J. Hendy at a cost of $75,000. The following January, Mrs. Abbie Krebs was elected president (and continued to hold that office until her death in 1924), and Henry Krebs became vice-president. In May 1902, Caspar Lumber acquired the White & Plummer Company, which included the store at Noyo and 1,000 acres of timberland on the Noyo River. The following month, it adopted a ten-hour day. On July 1, 1904, the Caspar & Hare Creek Railroad was valued at $250,000. In addition to four locomotives, it had fifty-eight cars and fifteen miles of track (from Caspar to the Noyo River). When the earthquake of April 16, 1906 destroyed the Jug Handle Creek trestle (160 feet high and 1,000 feet long), it was immediately rebuilt. A report of that period stated that the annual cost of maintaining the loading chute at Caspar amounted to $75,000.[29]

By 1912 the Caspar Lumber Company owned 80,000 acres of timberland. That year, the company gained control of the Redwood Manufacturers Company of Pittsburg, California, a firm organized in 1903 by the principal redwood manufacturers for the purpose of seasoning redwood lumber for shipment to the East. Ships transported redwood from the sawmills of Mendocino County and Humboldt County to Pittsburg, where the climate was excellent for drying. With the construction of the California Western Railroad from Fort Bragg to Willits in 1912 and the extension of the Northwestern Pacific Railroad to Humboldt County in 1914, shipments by vessel to Pittsburg ceased, except in the case of Caspar Lumber, which had no rail connection to that point.[30]

Abbie Krebs Wilkins died in 1924 at the age of eighty-two, and Casimer Jackson Wood took over as president of Caspar Lumber. On March 22, 1926, the company bought the Fleming timber tract for the sum of $600,000. A few years later, the Depression hit the lumber industry, and from June 1931 to May 1934, the Caspar Lumber Company mill was shut down. In the spring of 1936, it began using tractors to transport logs from its camps to the railroad. A strike shut down mill and railroad in December 1945. By that time, the railroad was thirty-five miles long. When the mill started up again, Caspar Lumber found the road had deteriorated to the point where returning it to good condition represented a prohibitive expense, and it decided to replace it with trucks.[31]

On November 18, 1955, the Caspar Lumber Company mill closed down for the last time, an event which had been anticipated for many months and was precipitated by the company's decision to sell its logs and lumber to the Union Lumber Company of Fort Bragg, following a log deck fire at the latter's mill. This transaction involved an estimated 2 million board feet of logs and 7 million board feet of lumber. Until its last day, the Caspar mill, never having been converted to electricity, operated on steam power.[32]

In 1947 the State of California began acquiring

The Caspar Lumber Company at the mouth of Caspar Creek. The first mill was built here in 1861, and in 1864 Jacob Green Jackson bought the mill, and the Caspar Lumber Company remained in the family until 1955 when the mill closed down. — NANNIE M. ESCOLA COLLECTION

The Albion Lumber Company mill at the mouth of the Albion River and the nearby community of Albion. Although the ownership of the mill changed hands through the years, this mill continued from the early 1850's to 1928. — NANNIE M. ESCOLA COLLECTION

the Caspar Lumber Company timberlands, completing its purchase of them in 1952. They now constitute the Jackson State Forest. Since it still owns considerable property, Caspar Lumber continues in existence. Although the mill has long since been razed, many of the company's buildings and houses are still in use. When he died on April 7, 1964, at the age of ninety-four, Casimer Jackson Wood was still president of his grandfather's lumber business.[33]

****

Early in the 1850's, Captain William Richardson, who owned 10,000 acres of timberland along the Albion River in Mendocino County, let out a contract to Hegenmeyer and Scharf for construction of a water-powered sawmill at the river's mouth. In 1854 Alexander Macpherson, Alexander Dallas, and Donald Davidson acquired Richardson's mill and timber and built a steam sawmill, equipped with a sash saw, which could turn out 4,000 board feet of lumber each day. In 1855 the addition of a single circular saw and a planer increased the mill's capacity to 14,000 board feet daily. The new owners found a ready market in San Francisco, and the town of Albion built up around their mill.[34]

On September 1, 1864, Henry Wetherbee joined Alexander Macpherson in creating the firm of Macpherson & Wetherbee to operate the Albion mill and also a mill on the Noyo River. (In 1869 these two men would also start The Pacific Lumber

Company in Humboldt County's Eel River Valley.) The Albion facility burned down in 1867, and Macpherson & Wetherbee rebuilt it the same year, equipping it with a double circular saw, a sash saw, two planers, a picket saw, and an edger — thus increasing its daily capacity to 35,000 board feet. On May 31, 1879, the partnership was dissolved, and Henry Wetherbee and Miles Standish took over both mill and timber holdings. In September of the same year, fire at Albion destroyed the new mill, the dry kiln, a hotel, a barn, and several houses. Losses totaled $150,000, but Wetherbee and Standish quickly rebuilt.[35]

Five years later, Henry Wetherbee incorporated the Albion River Railroad, which included a 200-foot bridge across the river, a mile above Tidewater Gulch, and later a branch line to Railroad Gulch.[36] The Albion Lumber Company, with $300,000 worth of authorized stock, was incorporated on May 26, 1891 by Miles Standish, Henry Hickey, George C. Wilcox, and W. E. Reed. Reed became its president. From Henry Wetherbee, Albion Lumber purchased all lands, sawmills, cattle, horses, mules, railroads, and the tug *Maggie* for the sum of $37,500. By this time, Wetherbee's Albion River Railroad had been gradually extended to Keene's Summit, a distance of eleven miles from its original terminus. By 1895 Albion Lumber had established a thriving community at the Albion River's mouth, a unique feature of which was the prohibition by the company of the sale of alcoholic beverages within the town's limits.[37]

Robert H. Swayne purchased the Albion River Railroad from the Albion Lumber Company in April 1902 for $67,500, and on May 8 of that year incorporated the Albion & Southeastern Railroad with an authorized capital stock of $1 million. Swayne planned to extend the railroad to Booneville and run a general railroad freight and passenger business from there to Albion and beyond, by vessel, to San Francisco and other distant points.[38]

G. X. Wendling formed the Wendling Redwood Shingle Company to build a sawmill on Soda Creek, about twenty miles inland from Albion, and on June 16, 1902 reached an agreement with the Albion & Southeastern for an extension of the railroad to his mill site. On April 15, 1905, A. G. Stearns incorporated the Stearns Lumber Company and purchased Wendling Redwood Shingle. In turn, the Navarro Lumber Company bought out Stearns Lumber in 1914.[39]

One of the largest transactions to occur in the Mendocino redwood region took place on August 31, 1907, when the Albion Lumber Company was sold to the Southern Pacific Company, which needed an unlimited supply of ties and timber for construction of a railroad in Mexico. The price, which was reported to have been somewhere between $800,000 and $1 million, included the steamer *Pasadena*; the logging railroad and locomotives; 20,622 acres of land with an estimated 374,521 board feet of timber; a two-band sawmill; a planing mill; seven dry kilns; an engine house; a combined office, store, and post office; the Albion Hotel; a cook house; a hospital; houses; cabins for single men; and a dock 879 feet in length.[40]

In 1920 the Southern Pacific was again in need of timber and ties for extending its Mexican railroad from Guaymas to Guadalajara, and on May 15 of that year, it bought 40,000 acres of timberland. On August 25, 1920, Southern Pacific purchased the Navarro Lumber Company for $247,750 and became the owner of its logging railroad, rolling stock, sawmill, hotel, lodging house, shops, and other structures.[41]

Once its Mexican line was completed, however, the mighty Southern Pacific no longer needed its lumber mills, which by then were in a state of disrepair. On September 30, 1927, therefore, the Albion Lumber Company closed down its Navarro mill (Wendling's), and the Albion mill cut its last log the following May. In 1940 Southern Pacific sold the machinery and buildings it had acquired from the Albion Lumber Company to a dealer in Oregon.[42]

When the Southern Pacific bought the Albion mill in 1907, Miles Standish and Henry Hickey retained most of their timber holdings, which amounted to about 40,000 acres of redwood, and then formed the Standish & Hickey Timber Company, which was tied in with eight other investment firms. Although Standish died in 1942, the organization they formed in 1892, the Albion Lumber Company, lasted until 1951. That year, the entire holdings of Standish & Hickey, Inc., which at that time included about 70,000 acres of timber, were sold to the Masonite Corporation for use in operating its huge plant at Ukiah.[43]

****

Mendocino County's Rockport Redwood Company began to take shape in 1877, when W. R. Miller built a double circular mill at Cottoneva Creek. In 1887 the Cottoneva Lumber Company, with Joseph Viles of Santa Rosa as the major stockholder, was organized to take over Miller's development projects in the area. It assumed control of the mill at Rockport (as the Cottoneva area was then called). Cottoneva Lumber sold its holdings of 30,000 acres of redwood to the Finkbine-Guild Lumber Company of Mississippi in 1925. In order to carry on operations at Rockport, Finkbine-

The Finkbine-Guild Lumber Company plant at Rockport in 1927. The Rockport Redwood Company, which took over in 1938, was sold to the Georgia-Pacific Corporation in 1968, and in 1973 the new Louisiana-Pacific Corporation acquired the Cloverdale operations. — OREGON HISTORICAL SOCIETY COLLECTION

Guild had borrowed $2 million. By way of improvements, the new company had cable moorings fastened to a rock island off shore, so that ships could come under the cables to be loaded, and constructed a new, all-electric sawmill.[44]

In 1928, unable to pay off a mortgage held by the Des Moines Savings Bank & Trust Company, Finkbine-Guild was forced to sell its property to the Southern Redwood Corporation. Ten years later, a new firm, the Rockport Redwood Company, under the direction of Ralph M. Rounds, took over the former Finkbine-Guild mill and property. Rockport Redwood's lumber was not transported by vessel, but taken by truck to Fort Bragg and shipped from there by rail.[45]

In September 1942, fire destroyed Rockport Redwood's mill, log supply, and 6 million board feet of sawed lumber. In order to keep going, the company leased the Juan Creek mill, just south of Rockport, and by operating double shifts managed to meet its sales commitments until a new mill at Rockport was ready to function. In 1948 Rockport Redwood built a seasoning yard and drying plant three miles south of Cloverdale on the Redwood Highway, as well as resaw and planing mills to finish lumber hauled in by truck from the coast. In August 1968, the Georgia-Pacific Corporation purchased all timberlands and facilities of Rockport Redwood, and in 1973 the new Louisiana-Pacific Corporation took over the Cloverdale operations.[46]

H.B. Tichinor & Company's Navarro mill at the mouth of the Navarro River about 1880. This mill was one of the few that loaded from a wharf instead of a chute, and the manufacture of railroad ties made up a major portion of its output. — NATIONAL MARITIME MUSEUM — SAN FRANCISCO

Chapter Notes

1. L. L. Palmer, *A History of Mendocino County* (San Francisco, 1880), 429.
2. David Warren Ryder, *Memories of the Mendocino Coast* (San Francisco, Taylor & Taylor 1948), 56-57.
3. Palmer, *History of Mendocino County*, 429.
4. C. A. Menefee, *Historical and Descriptive Sketch Book of Napa, Sonoma, Lake and Mendocino Counties* (Napa, California, 1873), 333.
5. *Wood and Iron* XI-4 (April 1889): 93 and XXV-5 (May 1896): 93.
6. *Wood and Iron* XXXVIII-3 (September 1902): 18.
7. *The Humboldt Times*, January 8, 1904; Howard Brett Melendy, "One Hundred Years of the Redwood Lumber Industry, 1850-1950" (Ph.D. dissertation, Stanford University, 1952), 257.
8. *Wood and Iron* XLV-1 (January 1906): 10.
9. *Pioneer Western Lumberman* LVII-1 (July 1912): 7, 9.
10. Jack McNairn and Jerry MacMullen, *Ships of the Redwood Coast* (Stanford: Stanford University Press, 1945), 30; Ryder, *Memories of the Mendocino Coast*, 58.
11. Ryder, *Memories of the Mendocino Coast*, 10-16.
12. Ibid., 31-32.
13. Ibid., 32-35; *Wood and Iron* VII-7 (July 1887): 7 and IX-3 (March 1888): 40.
14. Ryder, *Memories of the Mendocino Coast*, 38.
15. *Wood and Iron* XXVI-1 (July 1896): 1 and XXVIII-4 (October 1897): 138 and XL-3 (September 1903): 21.
16. Melendy, "One Hundred Years," 257.
17. *Ford Bragg Advocate*, October 17, 1923.
18. Ryder, *Memories of the Mendocino Coast*, 75-76.
19. *The Humboldt Standard*, February 2, 1940.
20. B. J. Vaughn, Boise Cascade, Union Lumber Region, private interview, Fort Bragg, California, March 5, 1970.
21. Palmer, *History of Mendocino County*, 433.
22. Stanley T. Borden, "Caspar Lumber Company," *The Western Railroader* 315-316 (1966): 3-17.
23. Palmer, *History of Mendocino County*, 262.
24. Ibid.; Borden, "Caspar Lumber Company."
25. *Fort Bragg Advocate*, May 29, 1889.
26. Borden, "Caspar Lumber Company."
27. *Wood and Iron* XXXV-5 (May 1901): 173.
28. Borden, "Caspar Lumber Company."
29. Ibid.
30. Ibid.
31. Ibid.
32. Ibid.
33. B. J. Vaughn, Boise Cascade, Union Lumber Region, private interview, Fort Bragg, California, March 25, 1970.
34. Stanley T. Borden, "The Albion Branch," *The Western Railroader* XXIV (December 1961): 3-11.
35. Ibid.
36. *The Humboldt Times*, February 28, 1880; Borden, "The Albion Branch."
37. Ibid.
38. Ibid.
39. Ibid.
40. *Fort Bragg Advocate*, September 25, 1907; Borden, "The Albion Branch."
41. Ibid.
42. Ibid.
43. *Wood and Iron* XLVIII-5 (November 1907): 13; *The Timberman* XLIII-3 (January 1942): 112; B. J. Vaughn, private interview, March 25, 1970.
44. Palmer, *History of Mendocino County*, 471; Melendy, "One Hundred Years," 240; *The Timberman* XXVI-10 (August 1925): 127; *Fort Bragg Advocate*, June 23, 1926.
45. Melendy, "One Hundred Years," 265; *Fort Bragg Advocate,* January 12, 1938.
46. *Fort Bragg Advocate*, July 14, 1943; *The Timberman* XLIX-4 (February 1948): 122 and XLIX-6 (April 1948): 122; Arthur Peterson, Louisiana-Pacific Corporation, private interview, Samoa, California, May 5, 1975.

A Hobbs, Wall & Company woods camp near Smith River about 1915. This company, which began operating in 1871, closed its Crescent City mill in 1939. — HENRY SORENSEN COLLECTION

Redwood Mills in Del Norte Country

Although Del Norte County contained less redwood than Humboldt or Mendocino Counties, what it did have contained a high percentage of clear lumber and was situated primarily on level ground, either along the Smith and Klamath rivers or on the plain surrounding Crescent City. Prior to 1869, four small sawmills produced lumber for local use, and during that year, the first shipment of redwood from Del Norte to outside parts left Crescent City. Subsequently, two mills dominated Del Norte County's redwood industry — the Crescent City Mill and Transportation Company and Hobbs, Wall & Company.[1]

****

On December 3, 1868, John Chaplin, Jacob Wenger, Dennis Tryon, B. P. Grant, and John Nickel formed the Del Norte Mill and Lumber Company. In addition to acquiring timber northeast of Lake Earl and near its mill site at the southern end of that lake, the company built a wooden track railway to a wharf in Crescent City. On May 16, 1869, the owners of Del Norte Mill and Lumber exchanged all their property and interests for certificates of stock in the Crescent City Mill

and Transportation Company (as Del Norte Mill and Lumber was re-named). A decade later, in 1879, Crescent City Mill and Transportation was owned by Wenger & Company, a firm that included A. M. Simpson of San Francisco; the Mendocino Lumber Company; and Jacob Wenger of Crescent City, one of the founders of Del Norte Mill and Lumber.[2]

On July 14, 1891, the Lake Earl mill burned to the ground in a fire that also consumed 500,000 board feet of lumber. Two years later, after dissolving his partnership with A. M. Simpson, Jacob Wenger joined forces with J. S. Kimball of Mendocino. They rebuilt the Lake Earl shingle band sawmill and appointed Byxbee and Clark of San Francisco (formerly the owners of the Navarro Mill Company in Mendocino County) selling agents for their Crescent City Mill and Transportation Company. J. S. Kimball left the firm in November 1895. By the following year, the Lake Earl mill was turning out from 50,000 to 60,000 board feet of good lumber daily. As cutting and felling moved farther away, logs were hauled to the mill on the Hobbs, Wall & Company Railroad. Finished lumber products continued to travel to Crescent City on the company's own railway.[3]

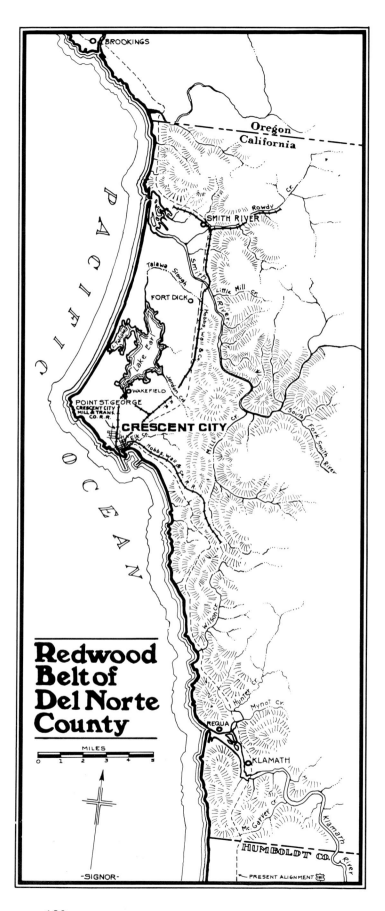

Redwood Belt of Del Norte County

MILES
0 1 2 3 4 5

-SIGNOR-

Jacob Wenger died in 1898, and his Crescent City Mill and Transportation interests passed to his former partner, A. M. Simpson. Simpson ran the company successfully until 1903, when he sold it to Hobbs, Wall & Company.[4]

****

Hobbs, Wall & Company had built a mill on Elk River, near the city limits of Crescent City, in 1871. The company's first owners were C. S. Hobbs, David Pomeroy, and General Joseph Wall. Hobbs and Pomeroy were partners in a San Francisco box factory, and General Wall had come to Crescent City in 1855. Originally, the company was named Hobbs, Pomeroy & Company. In 1879, however, Pomeroy was lost when the steamer *Mary*, on which he was returning to San Francisco from Crescent City, disappeared at sea, and the firm became Hobbs, Wall & Company. C. S. Hobbs died in San Francisco in 1875. His son, John K. C. Hobbs, succeeded him as president of Hobbs, Wall & Company and continued in that office until the company was sold in 1902.[5]

Hobbs, Wall & Company was a successful enterprise. In addition to the San Francisco box factory and its Elk River mill — a modern facility with an electric light plant — it owned two planing mills. Sawdust from both sawmill and planing mills was shipped to San Francisco and used as packing material in its boxes. By 1896 it also owned two logging camps, 4,800 acres of timber, and a 16-mile railroad. Together, camps, mills, and box factory employed 140 men. In 1899 the company tore down and redesigned its sawmill, in order to increase production, and the following year, it added a dry kiln and a shingle mill to its properties.[6]

Joseph C. Wall, the last of Hobbs, Wall & Company's first owners, died in Alameda on December 30, 1899, and on June 17, 1902, John K. C. Hobbs died at sea. Early in 1903, W. J. Hotchkiss of Port Blakely Lumber on Washington's Puget Sound purchased the company, but did not change its name. When A. M. Simpson sold the Crescent City Mill and Transportation Company to Hobbs, Wall & Company that year, the Hotchkiss interests gained control of the entire Del Norte County redwood industry, a power they would hold for the next forty years.[7]

In 1904 the officials of Hobbs, Wall & Company were W. J. Hotchkiss, president; D. E. Skinner, vice-president; and Louis Titus, attorney. In 1908 George M. Keller became the firm's Crescent City manager. Hobbs, Wall now owned and operated mills, camps, stores, steamers, and twelve miles of railroad from its camps on the Smith River to its

A Hobbs, Wall & Company locomotive No. 1, a 2-4-2 tank locomotive, bringing in logs to the mill at Crescent City about 1915. — HENRY SORENSEN COLLECTION

mills at Lake Earl and Crescent City. By 1920 the company owned 16,388 acres of Del Norte County timberland, one sawmill, a shingle mill, a general store, a loading wharf, and railroads. W. J. Hotchkiss of San Francisco, still its president, was also a director of eight other organizations. Hobbs, Wall & Company's main office was in San Francisco, and the capitalization and authorized amount outstanding of the company's stock was $1.5 million.[8]

In 1919 the Hotchkiss Redwood Company was formed to act as Hobbs, Wall & Company's timber-purchasing agent in Del Norte County. This firm foundered in 1927. In 1931 Hobbs, Wall defaulted on its bond, and a bond-holders' protective committee was appointed. From 1930 to 1934, during the Depression years, the company's mill operated only intermittently, and in 1935 the company filed a petition for reorganization under Section 77-B of the Bankruptcy Act, which gave creditors preferred stock for their claims against the firm. That same year, Howard A. Libbey, later president of the Arcata Redwood Company of Humboldt County, became Hobbs, Wall & Company's vice-president and general manager at Crescent City.[9]

During the next two years, misfortune continued to plague the company. Maritime strikes forced it to close down its operations for a time. Flood waters brought the machinery of its mill to a halt. A local teamster's strike prevented ships from being loaded. Labor troubles shut down the mill completely. The harbor filled with sand, again making it impossible to load vessels. After operating at a loss from 1936 to 1938, Hobbs, Wall & Company closed its Crescent City mill for the last time.[10] Its holdings were sold to A. M. King in 1940, and its successor, the North Coast Redwood Company, sold its entire lumber inventory. When the Hobbs, Wall & Company mill and machinery were dismantled in 1941, Crescent City and its surrounding area saw the end of its redwood industry.[11]

The Miller Redwood Company plant near Crescent City. This company, which was incorporated in 1954, is a subsidiary of the Stimson Lumber Company of Forest Grove, Oregon with offices in Portland. — DAVID SWANLUND

Although small sawmills existed in the Klamath area from the 1850's on, redwood and other lumber industries there did not prosper particularly until the 1940's, when logging trucks changed the transportation picture. The Klamath-Redwood Company was established in the 1920's, forced to close down by the Depression, and sold in 1945 to the Coast Redwood Company. In turn, in 1948, Coast Redwood sold its Klamath-Redwood holdings to the Simpson Logging Company of Washington. Simpson Logging subsequently began big-scale operations in Del Norte County, transporting rough lumber by truck from Klamath to Arcata, where it could be finished and distributed to the market.[12]

The two major redwood firms in the Crescent City area today are the Miller Redwood Company and the Simonson Lumber Company. Miller Redwood and the Rellim Redwood Company own the only old growth redwoods in Del Norte County. Their timberlands are approximately five miles southeast of Crescent City, in the Mill Creek drainage area, and were purchased by Mr. and Mrs. Harold Miller over a period of time from 1942 through 1944. Rellim Redwood was incorporated in California in 1954 and is a wholly-owned subsidiary of the Stimson Lumber Company of Forest Grove, Oregon, with main offices in Portland. Miller Redwood, incorporated in 1962, is also a wholly-owned subsidiary of Stimson Lumber, whose principal stockholders are members of the Miller family, which came to California from Michigan by way of Washington and Oregon.[13]

Rellim Redwood manages Stimson Lumber's Del Norte County timberlands, and Miller Redwood is responsible for the manufacture of products made from its logs. The capacity of the Miller Redwood sawmill, immediately to the southeast of Crescent City, is approximately 90,000 board feet of lumber per shift. About 280 men constitute the total work force in woods, mill, and veneer plant.[14]

The Simonson Lumber Company plant at Smith River. Arcata National purchased this company in 1979 for approximately $80 million. — DAVID SWANLUND

On July 2, 1979, Arcata National (ARCO) purchased the Simonson Lumber Company from its original owners for approximately $80 million. Simonson Lumber, a California corporation, belonged to Mr. and Mrs. L. H. Simonson and their children. L. H. Simonson began logging in the State of Washington in 1937 and moved to Del Norte County in 1950. Simonson Lumber, which operates on a sustained-yield basis, owns 70,000 acres of timberland in the Smith River area, some 35,000 acres of which are considered second growth redwood. Employing about 456 men, the company operatres both a sawmill and a stud mill in the town of Smith River. Together, these mills produce close to 375,000 board feet of lumber a day. Redwood constitutes roughly 100,000 board feet of the total.[15]

Chapter Notes

1. Howard Brett Melendy, "One Hundred Years of the Redwood Lumber Industry, 1850-1950" (Ph.D. dissertation, Stanford University, 1952), 118.
2. Ibid., 123; *San Francisco Bulletin*, June 6, 1879.
3. *Wood and Iron* XXII (October 1894): 140 and XXIV (November 1895): 178 and XXV (April 1896): 138 and XXVIII-2 (August 1897): 1.
4. *Wood and Iron* XXIX-4 (April 1898): 140 and XL-3 (September 1903): 21; *The Humboldt Times*, September 25, 1900.
5. *Wood and Iron* XL-3 (September 1903): 21; *Eureka West Coast Signal*, April 5, 1871; Melendy, "One Hundred Years," 128.
6. *San Francisco Bulletin*, June 6, 1899; Melendy, "One Hundred Years," 130; *Wood and Iron* XXVI-4 (October 1896): 141 and XXXV-6 (June 1901): 219.
7. *Wood and Iron* XXXIII-1 (January 1900): 10 and XXXVIII-1 (July 1902): 19 and XL-2 (August 1903): 17.
8. *The Humboldt Times*, June 18, 1908; Steve Scotton, *Del Norte County, California, Its Industries, Resources and Capabilities* (Crescent City, 1909), 11-12, cited by Melendy, "One Hundred Years," 134.
9. Melendy, "One Hundred Years," 135; *The Timberman* XXXVIII-2 (December 1935): 60; Howard A. Libbey, retired ARCO president, private interview, Eureka, California, May 22, 1972.
10. Melendy, "One Hundred Years," 138-140.
11. Ibid., 140; "Log and Saw, No. 115," *The Humboldt Standard*, April 11, 1949.
12. Melendy, "One Hundred Years," 142; "Log and Saw No. 115," *The Humboldt Standard*, April 11, 1949.
13. Darrell H. Schroeder, president, Miller Redwood Company, private interviews, Crescent City, California, February 2, 1970 and April 26, 1978.
14. Ibid.
15. *The Arcata Union*, July 5, 1979; Glenn Wallace, Simonson Lumber Company, private interview, Smith River, California, April 25, 1978.

The dedication of Lady Bird Johnson Grove at the new Redwood National Park on August 27, 1969. From left to right: Lyndon Johnson, Lady Bird, Patricia Nixon, President Richard Nixon, and Senator George Murphy. — DAVID SWANLUND

Impact of Redwood National Park

President Lyndon Johnson signed the Redwood National Park bill into law on October 2, 1968. Covering 58,000 acres in Humboldt and Del Norte counties, the park as described therein included 11,000 acres of virgin redwood. Its northern unit consisted of 20,852 acres, of which 5,625 acres were to be withdrawn from private ownership, and its southern units consisted of 34,717 acres, of which 22,476 were to be withdrawn from private ownership.[1]

Under this 1968 legislation, the Arcata Redwood Company lost seventy percent of its standing timber to the Redwood National Park. On June 29, 1971, however, the Arcata National Corporation of Menlo Park, Arcata Redwood's parent company, received 10,500 acres of Del Norte County timberland from the federal government by way of recompense for the holdings the company had given up. The Simpson Timber Company also lost timberland, but not enough to seriously curtail its operations. The Rellim Redwood Company, after an exchange with the government, lost only 2,650 acres.[2]

On March 27, 1978, after ten years of bitter controversy among conservationists, the timber industry, and organized labor, President Carter signed House Resolution 3913, expanding the Redwood National Park by 48,000 acres. Implementation of this bill will cost an estimated $359 million, plus $40 million for job protection and $35 million for park rehabilitation, bringing the total estimated cost close to $432 million. The protection package provides for retraining of loggers whose jobs are lost as a result of the legislation and an income guarantee of up to six years' duration for loggers who do not find suitable jobs upon relocation.[3]

Because of the expansion of the park's area, the Louisiana-Pacific Corporation lost about 27,000 acres of timberland, for which it received a payment from the government of $230 million. Arcata National parted with approximately 10,700 acres, for which the government gave it $80 million. The Simpson Timber Company lost 8,100 acres. Of the lumber companies involved, only Arcata National promised to reinvest the monies received in the northern Pacific coast area. Shortly after the 1978 park bill was signed, J. Frank Leach, president of Arcata National, made this public statement:

> We can concentrate our energies now on the fulfillment of our commitment to sustain our redwood operations and to maintain our posi-

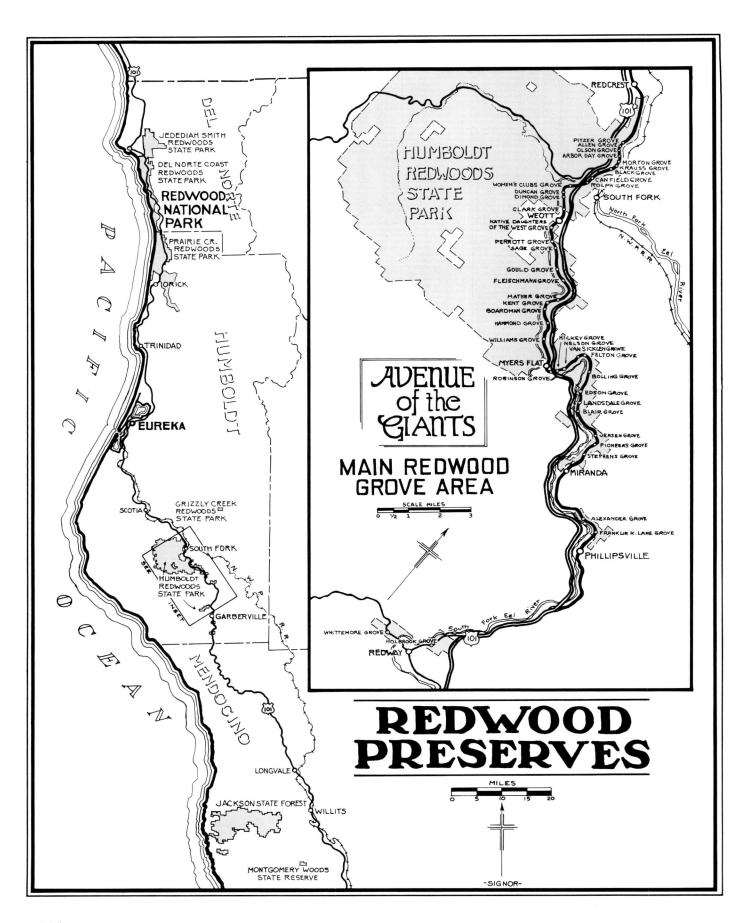

REDWOOD PRESERVES

Avenue of the Giants

MAIN REDWOOD GROVE AREA

tion as a strong supplier of specialized redwood and whitewood products. All money received from the federal government as a result of the expansion will be invested in qualified properties as rapidly as possible.[4]

The 1978 legislation amends and augments the earlier Redwood National Park act as follows:

1. It expands the area of the original Redwood National Park by 48,000 acres.
2. It provides for a 30,000 acre "Park Protection Zone" upstream from the park, and gives the Secretary of the Interior authority to acquire zone land.
3. It gives the federal government title to trees felled by man in the expansion area before January 1, 1975 or after January 31, 1978 and permits companies to remove other down trees, with the exception that removal of timber in streambeds requires the permission of the Secretary of the Interior.
4. It permits continued access, at existing levels, to acquired segments of roads. However, use of such roads is limited to forest and land management and protection purposes, including timber harvesting and road maintenance.
5. It stipulates that recovery of compensation for lands, trees, etc., taken will be handled by the United States District Court in San Francisco.
6. It directs the Secretary of the Interior to acquire lands for a by-pass highway around the eastern boundary of Prairie Creek Redwood State Park.
7. It authorizes, for protection and employment purposes, the sum of $33 million for federal government use in contracting for a program to rehabilitate areas within and upstream from the park that contribute sedimentation due to past logging and road building.
8. It directs the Secretaries of the Interior, Agriculture, Commerce, and Labor to analyze federal actions taken to mitigate adverse economic impacts on local communities (public and private sectors) and to consider contracting with local governments for forest resource improvement and utilization programs.
9. It directs the Secretary of Agriculture to present to Congress within one year of enactment a study of timber harvest scheduling alternatives for Six Rivers National Forest.
10. It provides for preferential hiring of employees affected by park expansion for federal government positions created by such expansion and for federal civilian jobs and by private employers whose undertakings involve federal participation, and it directs the Secretary of Labor to make training available to qualify affected employees for such jobs, and it guarantees industry that consideration or hiring of such employees will not subject employers to additional liability or obligation under equal employment laws.
11. It requires the Secretary of the Interior to submit to Congressional Committees by January 1, 1980, a comprehensive general management plan for the park and further requires an annual written report from the Secretary of the Interior to the Congress for a period of ten years.
12. It directs the Secretary of Labor to assure that the rights and benefits of affected employees, i.e., those totally or partially laid off between May 31, 1977 and September 30, 1980, are retained.
13. It provides for weekly income payments to such affected employees by the federal government, not to exceed previous earnings nor extend beyond September 30, 1984.
14. It provides for severance pay for up to seventy-two weeks from the federal government, not to exceed one week of pay for each month of service, and stipulates that acceptance of severance terminates other benefits.
15. It authorizes training of affected employees at federal government expense during the period of protection and a relocation allowance—both under specific conditions.[5]

Within two weeks of the passage of the 1978 park expansion bill, a number of small companies in the Humboldt area were forced to close. The list, together with the number of employees laid off by each, reads as follows: Trend Lumber Company Stud Mill, 175; Twin Parks Lumber Company, 60; Ramey Lath Mill, 15; McAllister Lath Mill, 11; and Gran-Star Redwood, 14.[6]

The Twin Parks Lumber Company of Arcata was forced out of business after thirty-one years because it could no longer get redwood logs from Simpson Timber. The other small remanufacturing plants depended on Louisiana-Pacific for most of their lumber and "edging" (the outside part of a redwood log), with which to make fencing and water-cooling slats used by oil refineries, atomic energy plants, and large building complexes. Park expansion made such demands on the big companies that they could not afford to sell more timber or lumber. As the availability of redwood decreased, its price climbed sharply, and the small plants simply could not afford to keep going.[7]

The McIntosh Lumber Company of Blue Lake, which began operating in 1950, laid off most of its men in November 1978, eight months after passage of the expansion bill, but continued to keep one shift going until September 1979, when it let its

(ABOVE) Patricia Nixon and Lyndon Johnson listen to President Richard Nixon speak at the dedication of Lady Bird Johnson Grove. (RIGHT) From left to right: Reverend Billy Graham, Governor Ronald Reagan, Congressman Don Clausen, Secretary of Interior Walter Hickel (at podium), President Nixon, Patricia Nixon, Lyndon Johnson, and far right, Senator George Murphy. — DAVID SWANLUND

remaining thirty employees go and auctioned off its machinery and other mill equipment. In December 1978, the Twin Harbors Stud Mill, the last in the Mad River-Dinsmore area of southern Humboldt County, closed down because of a shortage of timber, leaving some forty-four employees jobless.[8]

What the future holds for Humboldt County depends on the overall reactions of its three major forest products companies, the Louisiana-Pacific Corporation, the Arcata National Corporation, and the Simpson Timber Company. All three plan to convert some facilities to accommodate second growth redwood. Neither Arcata National nor Simpson Timber anticipate lay-offs, but Louisiana-Pacific does. Moreover, park expansion will have an impact on employment outside the lumber industry but dependent upon it, e.g., trucking firms and equipment manufacturers.[9]

Louisiana-Pacific engages in some trading of logs and is also the lumber supplier for numerous small local mills. L-P officials have declared that the company will not plow its government payment of $230 million back into timberland purchases, but will spend the money on production facilities. As a result, L-P employees are expected to be more severely affected than those of the other two big companies. For example, in the first week of April 1978, L-P altered its sawmill at Samoa to handle Douglas fir, rather than redwood, for one month. Since processing Douglas fir requires fewer men,

forty-nine employees were laid off.[10] On June 22, 1979, the L-P stud mill ceased sawing operations and forty-six employees were let go.[11] As of September 1979, the company was constructing a new mill to handle smaller logs in the former plywood building.[12]

Officials of Arcata National still insist that their company has every intention of reinvesting in the area, and that it is planning to build a fir mill. As proof of their commitment, they cite Arcata's January 1978 purchase of about 13,000 acres of predominantly fir land in Humboldt and Trinity counties from the A&M Timber Company. Although Arcata National's plant at Orick will operate on a single shift, instead of two, the management hopes to keep all its workers.[13]

The Redwood National Park expansion left Arcata National with approximately 45,000 acres of land on the northern Humboldt County coast, but the company is counting on the Forest Service allowing it more cuts. Six-Rivers National Forest alone contains almost 260,000 acres of land. According to the expansion bill, the Forest Service must have outlined the wilderness areas within one year of the President's signing the legislation. A replacement cut could considerably limit the impact of the expansion on Arcata National. Louisiana-Pacific, too, is preoccupied with the future of the wilderness areas, because it is counting on Forest Service cuts to keep its Carlotta mill going.[14]

On February 21, 1979, Arcata National announced plans to acquire the Simonson Lumber Company of Del Norte County, and the transaction was completed on July 2, 1979 for approximately $80 million. Included in the sale were two sawmills, a planing mill, and a nursery, all of which employed about 450 persons, as well as 30,000 acres of timberland (which increased the Arcata National holdings to roughly 76,000 acres). According to the company, the purchase gives the firm its first whitewood and fir mill and also cuts its acreage losses as a result of the park expansion.[15]

The Simpson Timber Company, least affected of the three major concerns, was gradually adjusting to less old growth redwood at the time of the expansion legislation and did not anticipate layoffs. In spite of its large existing land base, the company "is always in the market to buy land." Because of that base, however, it was not as concerned as L-P and Arcata National about the wilderness question. Although its officials foresaw no plant closures, they were planning a few changes, e.g., a gradual transition to handling more second growth logs at the Mad River plant and a future similar conversion of the Fairhaven and Korbel plants.[16]

On August 3, 1979, however, Simpson Timber announced it would lay off nearly 400 workers over the next two years, due to a "realignment" of the company's operations as it shifts emphasis from old growth to new growth. Most of the layoffs, it went on to say, would be in its plywood facilities, where approximately 315 jobs would be eliminated within five months. Thomas R. Ingham, Jr., Simpson Timber's vice-president in charge of California operations, informed the press that the company would (1) close its Mad River plywood plant at Arcata, laying of 295 employees; (2) undertake a modernization program at its Fairhaven plywood plant, cutting out 20 employees; and (3) carry out changes at its Korbel sawmill and other operations that would result in the elimination of another 80 jobs over the next two years.[17]

Chapter Notes

1. Michael McCloskey, "The Last Battle of the Redwoods," *The American West* VI (September 1969):64.
2. *The Times-Standard,* June 29, 1971.
3. *San Francisco Chronicle,* March 28, 1978; *The Times-Standard,* March 29, 1978.
4. *The Times-Standard,* March 29, 1978.
5. Karen Erickson, "Expanded Redwood Park Bill Is Signed," *Simpson Magazine* (April 1978): 9,11.
6. *The Times-Standard,* April 9, 1978.
7. Ibid.
8. Carl Dethlefs, private interview, Blue Lake, California, September 19, 1979.
9. *The Times-Standard,* April 9, 1978.
10. Ibid.
11. *The Arcata Union,* July 5, 1979.
12. Lois Bishop, Public Relations Officer, L-P, private interview, Samoa, California, September 13, 1979.
13. *The Times-Standard,* April 9, 1978.
14. Ibid.
15. *The Arcata Union,* July 5, 1979.
16. *The Times-Standard,* April 9, 1978.
17. *The Arcata Union,* August 9, 1979.

James Rydelius, Simpson Timber reforestation supervisor, checks over 2,000 redwood plantlets transplanted from the original test tubes to individual plastic containers and transferred to the Korbel greenhouse. — SIMPSON TIMBER COMPANY COLLECTION

The Promising Future

Over a decade ago, here and there in the Redwood Country, foresighted lumbermen, aware of how rapidly the supply of massive old growth timber was disappearing, began building mills designed to handle younger and smaller logs. Today, in the lumber development of the century, major redwood producers have joined the trend. Old redwood mill procedures, set up to process logs of ten feet or more in diameter, are becoming a thing of the past, and new mills with lighter and faster equipment cope smoothly with logs five or six inches in diameter from trees now turning up in steadily increasing numbers on privately held land. Recognizing in this a tremendous potential for industrial growth and strengthened economic stability, individual producers, as well as the Redwood Association, are actively and successfully seeking and creating markets for new products made from a new kind of timber, and the evolution of "small log" mills is favorably affecting everything from local job opportunities and tax revenues to long-term forest management planning on the part of the lumber industry.[1]

Mendocino and Sonoma county timberlands were the first redwood areas to be heavily cut over,

and the use of small logs began there; but the number of small log mills in the traditional old growth areas of Humboldt and Del Norte counties has steadily increased and will accelerate again with the actual expansion of the Redwood National Park. To compensate for their loss of land and old growth resources to government ownership, lumber companies will increase the small log yield of acreage they still own and invest in more second growth land.[2]

E. W. Thrasher built the first small log mill at Calpella (Mendocino County) in about 1964. Today, the Masonite Corporation owns and operates it as its Western Timber Division headquarters, and the company has also remodeled a plant at Cloverdale to handle small logs. In Humboldt County, Schmidbauer Lumber Company at Eureka manufactures pre-cut fencing and channel siding from small logs, while The Pacific Lumber Company's small log mill has been functioning in Fortuna since 1972 and Georgia-Pacific runs a new mill for small logs in Fort Bragg.[3]

Since small logs produce a higher proportion of knotty sapwood-streaked lumber than old growth, they require new markets, as well as new machinery. The California Redwood Association is pro-

Dr. Stanwood Schmidt

Dr. Stanwood Schmidt graduated from the University of California and later the University of Rochester Medical School. After five years of postgraduate training at the University of California in San Francisco, he came to Eureka in 1950 to establish his private practice in urology. He has authored over forty-five publications in medical journals and is an internationally recognized authority on male sterilization and its reversal. Because of his expertise in this field, he is a research associate in urology at the U.C.S.F. Medical School. Since coming to the Redwood Country, he has been interested in the history of the redwood lumber industry and has worked through the years with the State of California to establish the logging museum at Fort Humboldt in Eureka.

moting this lumber nationwide as "redwood garden grade," ideal for decks, fences, and garden shelters.[4]

Putting small logs to use brings about better utilization of all redwood resources. For example, the new mills consume logs from forest-thinning operations that stimulate the growth of big trees, thus making each acre more productive. The small log trend is also moving industry closer to sustained-yield forest management, for which redwood is particularly well-suited, since it is the only commercial species that sprouts new trees from cut-over stumps and also the fastest-growing commercial conifer in the United States.[5]

Already, The Pacific Lumber Company is 100 percent self-sufficient because of its sustained-yield management, while other major redwood producers are relying more and more heavily on harvesting and regrowth plans for managing their forests through the next century.

The natural corollary to these developments has been a growing demand for enormous quantities of reforestation stock. The Simpson Timber Company, for instance, has doubled the size of its forest nursery at Korbel, so that the facility now holds 9 million containerized seedlings. The addition of shade frames adjacent to the nursery buildings in 1977 made it possible to grow two crops of seedlings each year. Four months after the first crop of 4.5 million seedlings gets a start on accelerated growth in the controlled environment of the nursery's fifteen greenhouses, they are moved outside to the shade frames, where they toughen and become better able to withstand outdoor planting. As soon as the initial crop has been transferred outdoors, the second crop of 4.5 million is seeded into individual tubes, which go into the greenhouses. The infant trees are sold to the U. S. Forest Service, as well as to local and distant private landowners in California.[6]

On May 1, 1978, the Simpson Timber Company and the University of California at Irvine jointly announced a tremendous breakthrough in applying the science of genetics to redwood forestry. Using a tissue culture process, Dr. Ernest Ball, professor emeritus of biological sciences at the Irvine campus, had discovered that when a tiny portion of a mature redwood is placed in a test tube containing a specific combination of chemicals, it will grow shoots, needles, and roots. In ten months, such "super kid" redwoods, the offspring of blue ribbon parentage, will grow to six inches and be ready for planting in the forest.[7]

The breakthrough had come after three years of guarded research funded by the Simpson Timber Company and carried out by Dr. Ball, and its immense significance was that it could lead in the near future to the reforestation of Simpson Timber lands in Northern California with millions of offspring of the very best redwood species. Forester James Rydelius translated the discovery into concrete statistics:

> By using genetically superior planting stock, we can increase timber growth at least fifty percent an acre. The average yield per acre might leap from 50,000 to 75,000 board feet.[8]

The process of turning Dr. Ball's successful experiment into practical application is currently under way. Despite the impressive laboratory evidence, a number of questions must still be answered and numerous obstacles overcome before the commercial possibilities of tissue culture growth can be proven. Such questions and obstacles, as well as probable costs, are under study by Simpson Timber, which is evaluating the installation of a tissue culture pilot near its Korbel nursery as a first step toward the goal of reproducing all the redwood the industry can use.

Trees grown from tissue cultures are expected to show characteristics identical to those of the tree from which the original tissue was obtained. If Dr. Ball's process proves commercially feasible, the "super kid" progeny, branching from a superior family tree, will cause a significant reduction in the time now required to reproduce redwood timber, as well as an increase in product volume and quality. In the short time the project has been going on, fantastic progress has been made in producing plantlets, although not without the usual scientific frustrations.

Nevertheless, by continuous experimentation with the chemical medium or "agar" in which the plantlets are developed, Dr. Ball and his assistant, Dawn Morris, are now producing healthy young trees that grow as the theory says they should. Rydelius has 2,000 of the new breed in his own greenhouse, of better quality than seedlings of the same age in the Korbel nursery, which will be planted on Simpson Timber Company land. The basic material for these "super kids" came from about 200 trees in Simpson Timber's young growth forest. Chosen from an estimated total of a million trees examined, they show evidence of rapid growth and have better-than-average straight limbless trunks.[9]

There remains to be told only the story of how this giant step toward the renewal of *los palos colorados* came to be taken. Almost ten years ago now, a Eureka physician, logging buff, and amateur botanist by the name of Dr. Stanwood Schmidt approached Hank Trobitz, Simpson Timber's California Resources Manager, and asked him if he would like to have 100,000 Howard Libbey trees, because—he said—there had been a

The Simpson Timber Company nursery at Korbel holds nine million containerized seedlings. — SIMPSON TIMBER COMPANY COLLECTION

Dr. Ernest Ball

Dr. Ernest Ball examines one of his plantlets at the Korbel nursery. The "super kid" progeny is expected to cause a reduction in time required to reproduce redwood timber, as well as an increase in product volume and quality. — SIMPSON TIMBER COMPANY COLLECTION

technique called "Meriston culture" going on in orchid growing for the past twenty years that permitted a rapid production of thousands of plants, and it would work with redwoods, too, since they sent up stump sprouts. Trobitz was persistently skeptical; but Dr. Schmidt kept on prodding him until he asked Forester Rydelius to look into a research project "just to get the Doc off my back!"[10]

Rydelius remembered reading a paper some years before by a Dr. Ball, who had published the results of some tissue culture research on *Sequoia sempervirens* while teaching at North Carolina State University in 1950, but who apparently had not done much with redwood since then. He tracked him down at Irvine. The first tissue culture research contract with Dr. Ball was signed in 1975, and Simpson Timber has invested $110,000 in the enterprise.[11]

The success of the project thus far rests in no small measure on the redwood's extraordinary ability to reproduce itself from stump sprouts, as well as seed—a characteristic of the coastal redwoods contemplated no less thoughtfully by Dr. Schmidt, wanting to insure their future, than their texture and color were contemplated by Fray Juan Crespí, wanting to give them a name, two centuries earlier near Monterey.

Chapter Notes

1. *The Times-Standard,* November 27, 1977.
2. Ibid.
3. Ibid.
4. Ibid.
5. Ibid.
6. Karen Erickson, "Reforestation," *Simpson Magazine* (January 1978):4.
7. Karen Erickson, "Test-Tube Forest with Bull-Team Roots," *Simpson Magazine* (April 1978): 3.
8. *The Times-Standard,* May 2, 1978.
9. Erickson, "Test-Tube Forest."
10. Ibid.; Dr. Stanwood Schmidt, private interview, Eureka, California, May 9, 1978.
11. Henry K. Trobitz, Simpson Timber Company, private interview, December 6, 1980.

CALIFORNIA *REDWOOD*
The World's most durable Lumber

STILL SOUND

Appendix

APPENDIX A — Early Humboldt County Lumber Mills

Early Eureka Mills in 1854 as painted by a soldier from Fort Humboldt. From left to right: the Smiley Mill, the Muley (Mula) Mill, and the Ryan and Duff Mill. — HUMBOLDT COUNTY HISTORICAL SOCIETY COLLECTION

Pioneer Mill, 1850: Supposedly the first mill established in Humboldt County. Built on Humboldt Bay by J. M. Eddy and Martin White, at what is now the intersection of Second and M streets in Eureka, and supplied with timber from woods operations between Ryan Slough and Freshwater. According to one documented account, Martin White then erected a second mill near Bucksport, called the Papoose Mill, which was in operation by November 1850. A differing version states that the Pioneer Mill was also known as the Papoose Mill. In 1857 the Pioneer Mill was sold to John Dolbeer at a sheriff's sale for the amount of its delinquent taxes.

Luffenholz Mill, 1851: Erected at Little River by Carl Luffenholz, who emigrated from Germany. The mill had been producing lumber for only a few years when it was destroyed by a freshet in 1854.

Modena Mill, 1852: Constructed at Bucksport and powered by machinery from the steamship *Chesapeake*. In 1855 the mill was abandoned and subsequently it was sold at a sheriff's sale.

Ryan & Duff Company, 1852: The steamship *Santa Clara* was run ashore at what is now the intersection of First and D streets in Eureka and converted into the first successful mill on Humboldt Bay, supplied with timber from woods operations near Freshwater. In 1858 Duff became sole owner and spent $11,000 rebuilding the mill, which had four gang saws and employed seventeen men. Before it was destroyed by fire in 1862, the mill was producing 60,000 board feet of lumber per twelve-hour day and employing thirty-five men.

Deming, & Marsh Company, 1852: Constructed a mill at Trinidad, which closed down a few years later.

Mula Mill, 1853: Began operating at what is now the foot of I Street in Eureka. Known as the Steymest and Company Mill during 1853 and 1854, it burned down in 1863.

Bay Mill, 1853: Constructed by Martin White, John Dolbeer, Isaac Upton, Dan Packard, and C. W. Long. One of the most successful mills in the redwood industry, it was operated by Dolbeer and McLean alone from 1856 to 1863. Destroyed by fire in 1860, it was immediately rebuilt and began operating again in two months' time.

Picayune Mill, 1853: Built by Hinkle, Flanders & Company at what is now the foot of K Street in Eureka. Sold to Charles McLean and John Dolbeer, owners of the Bay Mill, in 1860, and abandoned in 1861.

Vance Mill, 1853: Originally owned by Ridgeway and Flanders and situated at what is now the foot of G Street in Eureka. In 1854 the Vance Mill was the second largest on Humboldt Bay and contained two gang saws, a single saw, one edger, one crosscut saw, two lath machines, and one planer. It produced 50,000 board feet of lumber and 32,000 laths per day and employed thirty-one men.

Smiley or Bean Mill, 1854: Owned by Hiram Bean and J. C. Smiley and situated at what is now the foot of H Street in Eureka. In 1855 Smiley, by then the sole owner of the mill, which cut 20,000 board feet of lumber per day and employed eight men, sold a half interest to Kenelly and Long. In 1860 D. R. Jones and John Kentfield of San Mateo purchased the facility.

Dolbeer & Carson Lumber Company, 1864: Formed after Charles McLean, Dolbeer's partner, drowned in the *Merrimac* disaster of 1863. William Carson purchased a half interest in the Bay Mill, which Dolbeer and McLean had been operating. The Dolbeer & Carson Lumber Company lasted until 1950, when The Pacific Lumber Company bought both the Bay Mill and Dolbeer & Carson's timber holdings. Subsequently, the Simpson Timber Company took over the Bay Mill and established a fir plant at that location. In 1960 the Park Loading Corporation acquired the Bay Mill property.

D. R. Jones & Company, 1865: Formed when Captain H. H. Buhne joined Jones and Kentfield in operating the former Smiley Mill at what is now the foot of H Street in Eureka. The mill subsequently changed hands several times before it was destroyed by fire in 1898. In 1866 D. R. Jones & Company also operated the largest mill on Humboldt Bay at Gunther Island, a double mill with forty-six simultaneously operating saws that employed sixty-five men and turned out 70,000 board feet of lumber per day. Woods operations were in the Elk River and Salmon Creek areas. In 1884 D. R. Jones & Company was purchased by the California Redwood Company. When California Redwood suspended operations two years later, the Gunther Island mill was dismantled. In 1890 its machinery was moved to the Consumers Lumber Company, south of Samoa on the peninsula.

Occidental Mill, 1868: Built by William Duff on what is now First Street between A and B streets in Eureka. In December 1869, this mill burned down, but was promptly rebuilt. After its original owner went into bankruptcy in 1871, the mill was

John Talbot Ryan

James Talbot Ryan, the founder of Eureka, was a partner in the Ryan & Duff Company mill, the first successful mill on Humboldt Bay. He sold out his interest to Duff in 1858. The mill was destroyed by fire in 1862. — MARTHA ROSCOE COLLECTION

The Occidental mill, which was built by William Duff, operated from 1868 to 1932. The mill was destroyed by fire in 1934. The company had timber on Salmon Creek, Elk River, Ryans Slough, and Worthington Prairie. — REED STROPE COLLECTION

sold to Evans & Company and re-named the Occidental Mill. It had circular saws on the head rigs and cut 50,000 to 60,000 board feet of lumber per day. In 1875, when David Evans, A. Stinchfield, Flanigan, and Brosnan withdrew from Evans & Company, the Evans and Stinchfield interests were purchased by Allan McKay. When McKay died, in 1888, James Loggie, who had become associated with him through marriage, took over his business in Eureka and San Francisco, in which the Charles Nelson Steamship Company also held a large interest. The Occidental Mill operated until 1932 and was destroyed by fire in 1934.

Cousins Mill, 1869: Erected on Gunther Island by J. Russ & Company, which included Joseph Russ, Euphronius Cousins, and Wood. Woods' operations were at Elk River, and by 1875 the mill employed fifty men and was running day and night, with an average daily output of 45,000 board feet of lumber. In 1883 the mill was sold to the California Redwood Company, and in 1886 it became the Excelsior Redwood Company's mill, ultimately closing down in 1893, when Freshwater area timber had been logged off.

Dougherty and Smith Mill, 1869: Built by Dougherty and Smith, this was the first big mill in the Trinidad area. In 1873 it was consolidated with the Hooper Mill, owned by the Trinidad Mill Company.

Trinidad Mill Company, 1869: Formed by the Hooper brothers, who operated the Hooper Mill in Trinidad and, later, the Dougherty and Smith Mill in the same area. The company included F. T. Hooper, J. A. Hooper, and J. C. Smith, all of San Francisco. Josiah Bell of Trinidad served as a director and on-the-job superintendent. Sold to the short-lived California Redwood Company in 1883, the Trinidad Mill Company's holdings subsequently became the property of the Excelsior Redwood Company, whose major owners were the Hooper brothers, Charles and George. The Trinidad facilities burned down in 1886, ending the lumber industry in that particular area of Humboldt County.

Janes Creek Mill, 1869: Built by Noah Falk, leading partner in Falk, Chandler & Company, near Camp Curtis, just north of Arcata. In 1888, after all timber in the area had been logged off, this mill closed down. In 1890 it was destroyed by fire, and its machinery was later transferred to the Elk River Mill.

The Trinidad Mill Company sawmill on Mill Creek which was operated by the Hooper brothers of San Francisco. A tramway connected the mill to the wharf at Trinidad Head. — KATIE BOYLE COLLECTION

The Trinidad Mill Company locomotive, the *Sequoia,* with a load of logs for the mill. The engine was later sold to the Bucksport & Elk River Railroad. — CLARKE MUSEUM COLLECTION

The Janes Creek mill, which was built by Noah Falk just north of Arcata in 1869, lasted until 1888. — SUSIE FOUNTAIN COLLECTION

The Pacific Lumber Company, 1869: Formed chiefly by A. Macpherson and H. Wetherbee of the Albion Lumber Company (Mendocino County). Pacific Lumber owned property, but did no logging. A new company purchased its holdings in 1886 and began operating a mill at Forestville (now Scotia). Over the years, Pacific Lumber went through a succession of changes, and today it operates the largest redwood mill in the country.

Jolly Giant Mill, 1874: Built by Noah Falk and Isaac Minor and situated in a gulch just behind the present site of Humboldt State University. In 1875, when the mill was producing 25,000 board feet of lumber per day, Isaac Minor sold his interest in the Jolly Giant Mill and the Dolly Varden Mill to George Harrington, G. W. Chandler, and J. Hawley. Both mills, which together employed fifty men, were considered the best of their size in Humboldt County at that time. In 1885 the Jolly Giant Mill was sold, and in 1886 it was dismantled and its machinery moved to Trinidad for use in a shingle mill.

Big Bonanza Mill, 1874: Double circular sawmill built by John Vance at Essex, on the north bank of the Mad River, at the mouth of Lindsay Creek. In 1892 the Big Bonanza Mill's machinery was moved to the peninsula for installation in a new mill built by Vance's nephew and sons.

Milford Mill, 1875: Constructed by David Evans, John McKay, and H. A. Marks on Salmon Creek. This mill had double circular saws and could cut 25,000 board feet of lumber per day. The South Bay Railroad Company provided it with a logging road, as well as an outlet for its finished products. In 1876 the mill was purchased by William Carson, who sold it in 1902.

Flanigan, Brosnan and Company, 1876: Established on the west side of Eureka at the foot of Whipple Street by Dave Flanigan and Timothy Brosnan of Eureka and John Harpst and James Gannon of Arcata, who at the time also operated the Jolly Giant Mill and the Union Mill in Arcata. The founders later sold a five-sixteenths interest in the company to James Tyson of the Charles Nelson Steamship Company. In 1902 a new company took over the mill and re-named it the Bayside Mill & Lumber Company.

Warren Creek Mill, 1881: Double circular sawmill—the first to cut the magnificent timber that bordered the Mad River—constructed on the south bank of the Mad River at Warren Creek by Isaac Minor, Isaac Cullberg, and Kirk and enlarged in 1883.

The Jolly Giant mill about 1876, which was owned by Isaac Minor and Noah Falk. — A.W. ERICSON

Isaac Minor

Isaac Minor, with Noah Falk, built the Dolly Varden and the Jolly Giant mills. Later, in the Mad River area, he built the Warren Creek mill and the Glendale mill.

John Vance's Big Bonanza mill at Essex on Mad River. The Vance home is on the right. — HUMBOLDT COUNTY HISTORICAL SOCIETY COLLECTION

A panoramic view of the Elk River Mill & Lumber Company. The mill began operating in 1885 and closed in 1938. The first band saw headrig in Humboldt County was installed in this mill in 1888. — REED STROPE COLLECTION

The Humboldt Lumber Mill Company and log pond in 1900. The three Korbel brothers of Sonoma bought timber on the north fork of Mad River and built the town of Korbel. The brothers later sold all of their Mad River properties in 1902 and left the area. — HUMBOLDT COUNTY HISTORICAL SOCIETY COLLECTION

Before a railroad was built to the mill, logs for it were floated six miles down the Mad River to a canal that cut into Mad River Slough, and then one mile farther to the bay. In 1886 Isaac Minor bought out Cullberg and Kirk, and a year later, the mill was converted into a band sawmill. From fifty to fifty-eight men were employed in woods operations, and another thirty-eight men worked in the mill. Rebuilt after it was destroyed by fire in 1896, the mill burned down again in 1902 and was subsequently abandoned.

Elk River Mill and Lumber Company, 1882: Formed by Noah Falk, Irwin Harpster, C. Stafford, J. Hawley, and Ben Pendleton to operate its mill at Elk River (now Falk). The Elk River Railroad Company extended a road to the new mill, which began operating in 1885 and cut 40,000 board feet of lumber per day. In 1888 the company installed the first bandmill head rig in Humboldt County. When fire completely destroyed the mill in 1890, Noah Falk rebuilt it and equipped it with machinery from the Janes Creek Mill, which had closed two years before. In 1904 the J. R. Hanify interests of San Francisco became the major stockholder in the company, and the mill continued to operate successfully until it was closed down in 1938.

Chandler, Henderson & Company, 1882: Formed by C. W. Chandler, M. F. Henderson, A. Kendall, and F. Graham to operate a mill at Blue Lake, which by 1883 was producing 40,000 board feet of lumber per day. In 1886 the company closed the mill and moved its equipment across the river to Riverside, near Korbel, closer to its timberlands. In 1888 the original partnership was dissolved, and a new corporation, the Riverside Lumber Company, was established with Harry Jackson as president and general manager and F. Graham as vice-president and logging superintendent.

The California Redwood Company, 1883: Formed to buy up all mills and timber in the Humboldt region. California Redwood purchased D. R. Jones & Company's holdings, J. Russ & Company's Cousins Mill on Gunther Island, the Hooper Mill at Trinidad, and 100,000 acres of redwood timber. Subsequently, the company was forced to suspend operations as the result of government investigations into its fraudulent dealings.

The Humboldt Lumber Mill Company, 1883: Formed by the three Korbel brothers—Antone, Frank, and Joseph—of Sonoma, who purchased timber on the north fork of the Mad River and built a town near Blue Lake, which they named Korbel. When its mill, the first in the region to use a dry kiln, burned in 1886, the company, headquartered in San Francisco, rebuilt it at once.

The Eel River Valley Lumber Company, 1884: Formed by E. Dodge, Euphronius Cousins, and H. D. Cousins to operate a mill on Strong Creek, east of Fortuna, at Newburg. In 1887 Thomas Pollard of San Francisco bought an interest in the firm. In 1931, after all its timber had been cut, E. Dodge, who owned the company at the time, closed down the mill.

Glendale Mill, 1885: Built by Isaac Minor, James Kirk and Company on Hall Creek, near Blue Lake. When Minor retired, in 1895, his children formed the Minor Mill and Lumber Company, which in 1903 signed a contract with The Eastern Redwood Company under which the latter furnished logs for the Glendale Mill. The mill shipped lumber from Arcata to San Diego, from where it was transported to the East. In 1911, after all available timber had been logged, the Glendale Mill closed, and in 1915 the company was dissolved by court order.

The Excelsior Redwood Company, 1886: Formed by Charles and George Hooper of San Francisco and Joseph Russ and David Evans of Eureka after the California Redwood Company was forced to suspend operations. The former Cousins Mill on Gunther Island (erected by J. Russ & Company in 1869 and later sold to California Redwood) now became the Excelsior Mill and employed 300 men. In 1893, when its timber near Freshwater was logged out, the company closed down the mill, and in 1903 it was dismantled and shipped to Hardy Creek (Mendocino County), where the Hooper brothers were building a new mill.

John Vance Mill & Lumber Company, 1892: Formed to carry on the business interests of John Vance by his nephew John M. Vance. When the old Vance Mill at the foot of G Street in Eureka burned down in 1892, John M. Vance bought land on the peninsula, where he built the Samoa Mill, equipping it with machinery moved from the Big Bonanza Mill (also a Vance enterprise) at Lindsay Creek. In 1900 Vance Mill & Lumber was sold to A. B. Hammond, and in 1912 it became the Hammond Lumber Company. The Georgia-Pacific Corporation of Atlanta, Georgia purchased Hammond Lumber in 1956, and in 1973, when the U. S. Government forced the huge Georgia corporation to split into two separate entities, the spun-off

The Eel River Valley Lumber Company mill just east of Fortuna. Frank Hagman (in foreground) drowned in the log pond in 1903. — EVELYN MCCORMICK COLLECTION

A view of the Glendale mill near Blue Lake in 1890. Isaac Minor built the mill in 1885 and it closed down in 1911. — A.W. ERICSON

Drying yards and the Excelsior Redwood Company mill on Gunther Island in 1890. — A.W. ERICSON

A panoramic view of The Little River Redwood Company sawmill and the town of Bulwinkle in 1914. The company, which began operating in 1908, merged with the Hammond Lumber Company in 1931. — CLARKE MUSEUM COLLECTION

company, the Louisiana-Pacific Corporation, took over the Samoa facility and G-P's Humboldt County timberlands.

The Little River Redwood Company, 1893: Based in Tonawanda, New York, with principal owners in Ottawa, Canada and western New York State. Little River Redwood constructed a mill at Crannell near Little River and commenced operations there in 1908. In 1931 it merged with the Hammond Lumber Company and became the Hammond-Little River Redwood Company, Ltd.

The Bayside Mill and Lumber Company, 1902: Formed in 1902 to continue operating the mill established by Flanigan, Brosnan and Company in 1876 on the west side of Eureka. In 1905 the Warren Timber Company of Pennsylvania purchased Bayside Mill and Lumber. Later, the mill was sold to the Bayside Redwood Company. In 1929 the Dessert Redwood Company, which held timberlands on the Van Duzen River and Yager Creek, bought the mill and formed the Humboldt Lumber Company. In 1937 the mill was sold to the Hammond Lumber Company and became its Plant Two. In 1962, Georgia-Pacific sold the property to the Park Loading Corporation.

The Holmes-Eureka Lumber Company, just south of Eureka on Humboldt Bay, was incorporated by J.H. Holmes and J.R. Lane. The mill began operating in 1904 and lasted until 1959. — FREEMAN ART PHOTO — HUMBOLDT COUNTY HISTORICAL SOCIETY COLLECTION

The California Barrel Company plant at Arcata as it looked from 1925 to 1930. — JAMES LUNDBERG COLLECTION

The Northern Redwood Lumber Company and company town of Korbel in the early 1900's. — A.W. ERICSON

The Riverside Lumber Company on Mad River near Korbel in 1890. This mill and the Korbel interests merged to form the Northern Redwood Lumber Company in 1902. — A.W. ERICSON

The Northern Redwood Lumber Company, 1902: A merger of the Riverside Mill (see Chandler, Henderson & Company) and the Korbel interests (see Humboldt Lumber Mill Company) to which the Korbel brothers sold all their Mad River area properties. The Charles Nelson Steamship Company controlled a good deal of Northern Redwood's stock. The mill closed down in 1933, but resumed operation in 1942. In 1956 the Simpson Logging Company of Washington purchased both mill and holdings.

The California Barrel Company, Ltd., 1902: San Francisco manufacturers of barrels, wire-bound boxes, and wooden containers, who built a plant at Arcata that supported that community's economy for more than fifty years. In 1956 its Arcata facility had 1,125 employees on a 29-acre site in the southwestern part of the town. Over the years, California Barrel operated bolt camps at Long Prairie Creek, near the north fork of the Mad River; at Essex; at the head of Strawberry Creek; and at Dows Prairie. During the 1930's and 1940's, it purchased logs from Hammond Lumber. The paper industry gradually took over the box business, and in 1956 California Barrel sold its Arcata plant to Roddis Plywood for $12 million. In 1961 Roddis Plywood sold it to Weyerhauser, and in 1965 Weyerhaeuser sold it to the Arcata Redwood Company.

The Holmes-Eureka Lumber Company, 1903: Incorporated by J. H. Holmes and J. R. Lane. Holmes-Eureka Lumber bought frontage land on Humboldt Bay, just north of Bucksport. The mill it built there began operating in 1904 and could turn out 70,000 board feet of lumber per day. Holmes, who moved to San Francisco in 1910 to establish a wider market for its redwood product, died in 1939. On January 2, 1959, The Pacific Lumber Company took over the mill.

The Metropolitan Redwood Lumber Company, 1904: Formed by a group of men from Michigan and Wisconsin led by Tom Atkinson. Metropolitan Redwood bought 3,600 acres of redwood timber in the Slater Creek drainage and erected a mill on McDairmid Prairie, three miles below Scotia on the Eel River. It also owned about twenty-five homes, a hotel, and a store. When the timber had been logged out, around 1925, the mill closed down. In 1932 it was destroyed by fire, and by 1937 most of the homes had been moved to Rio Dell. Over the years, the company made a net profit of $5 million.

APPENDIX B — Early Mendocino County Lumber Mills

Mendocino Lumber Company, 1852: Originally the California Lumber Company, formed by E. C. Williams, Jerome Ford, Henry Meiggs, and Captain David Lansing, all New Englanders, who built the first sawmill on the Mendocino coast at Big River. In 1853 the company built another mill on the flat near the mouth of the river, and in 1857 the first mill was dismantled. Later, the company changed its name to the Mendocino Lumber Company, and in 1873 its mill was the most important one in the county. In 1906 C. R. Johnson of the Union Lumber Company and C. J. Wood of the Caspar Lumber Company took over Mendocino Lumber. Closed in 1931, the mill reopened for a short period in 1938 to cut logs that had been salvaged when a log raft broke up off the coast, after which it was dismantled.

Richardson Mill, 1852: Built by George Hegenmeyer for Captain William Richardson. This was the first mill constructed on the Noyo River. It began operating in 1853, but was carried out to sea by a freshet in 1854.

Albion Lumber Company, 1853: This company's first mill was originally built for Captain William Richardson, but by 1856 it was owned by Alexander Macpherson, who turned it into a steam mill. When the mill burned in 1867, Macpherson and his new partner, Henry Wetherbee, rebuilt it immediately. They equipped the new facility with a double circular saw, one sash saw, two planers, a picket machine, a shingle machine, and a lath machine, and it could cut 35,000 board feet of lumber per day. In 1891 Henry and Ellen Wetherbee incorporated the Albion Lumber Company, headquartered in San Francisco. In 1892 they sold Albion Lumber, which consisted of the mill and 20,000 acres of timberland, to the Standish-Hickey interests of Michigan for about $250,000. In 1895 the new owners laid out a community at the mouth of the Albion River that grew into a prosperous lumber town. They also leased land at the foot of Sixth Street in San Francisco and constructed a planing mill and dryer there. From San Francisco, Albion Lumber's products were shipped by rail to all parts of the country. The company was sold to E. H. Harriman's Southern Pacific Company in 1907, and the Albion Mill continued operating until 1929.

H. B. Tichenor & Company, 1861: Constructed a mill at the mouth of the Navarro River that could cut 10,000 board feet of lumber per day. When H. B. Tichenor died, in 1883, the mill was capable of cutting 35,000 board feet of lumber per day, and the company owned 30,000 acres of timberland. His surviving partner, R. G. Byxbee, carried on. The Navarro Mill was one of the few that loaded from a wharf, instead of a chute, and the manufacture of railroad ties constituted a major portion of its activity. In 1886 R. G. Byxbee took in Joseph Clark as a partner, and together they formed the Navarro Mill Company, which replaced H. B. Tichenor & Company. When their mill was destroyed by fire in 1890, they immediately constructed a new band mill. In 1893, with some $500,000 in debts outstanding, the Navarro Mill Company became insolvent. At the ensuing sheriff's sale, held in San Francisco, R. G. Byxbee, its co-owner, purchased the Navarro Mill Company's properties for $13,000 and assumed its $350,000 mortgage, but the new mill remained idle. When it, too, burned down, in 1902, the total loss was estimated at $60,000. In 1903 the Pacific Coast Redwood Company was formed by eastern and local capitalists to take over the Navarro Mill Company. They transferred $600,000 to the Anglo-California Bank in exchange for the bank's interest in Navarro Mill's holdings, which included 30,000 acres of timberland, the town of Navarro with its

A steam schooner waiting to load at the Albion landing.
— NATIONAL MARITIME MUSEUM — SAN FRANCISCO

wharves, and fourteen miles of railroad. In 1905 A. G. Stearns incorporated the Stearns Lumber Company, which bought out the Wendling Redwood Shingle Company. Wendling Redwood Shingle's mill, built by G. X. Wendling, was located twenty miles inland from Albion and one mile north of the town of Wendling. Originally called Wendling's mill, it now became known as Stearns' Mill. In 1914 the Navarro Lumber Company formed with R. T. Buzard as president and bought out Stearns Lumber, and the mill's name changed again—this time to the Navarro Mill. Eventually, this Navarro Mill and the town of Wendling became the town of Navarro, while the Navarro Mill Company's site (originally H. B. Tichenor & Company's, where the mill had burned down in 1902) became known as Old Navarro or Navarro Ridge. In May 1920, the Albion Lumber Company, by then a subsidiary of the Southern Pacific Company, purchased 40,035 acres of land and timber rights along the Northwestern Pacific Railroad from the Pacific Coast Redwood Company for $2.5 million, and in August of the same year, it purchased the Navarro Lumber Company for $247,750. In September 1927, Albion Lumber closed down the Navarro Lumber Company Mill.

Caspar Lumber Company, 1861: Kelley and Randall consructed a mill at the mouth of the Caspar River that could turn out 15,000 board feet of lumber per day. It was purchased in 1863 by J. G. Jackson. In 1880, when the mill was capable of producing 45,000 board feet of lumber per day, Jackson incorporated his holdings as the Caspar Lumber Company, When the original mill was destroyed by fire in 1889, Caspar Lumber built a new one, and in 1893 the company bought a tract of timber on the south fork of the Noyo River. In succeeding years, Jackson's heirs continued to run the firm, and by 1912 Caspar Lumber owned 80,000 acres of timberland. Except for the period from June 1931 to May 1934, when it was forced to close down as a result of the Depression, the mill remained in operation until November 1955. Caspar Lumber Company's timberlands ultimately became the Jackson State Forest.

Gualala Mill Company, 1862: A steam mill built on the Gualala River by Rutherford and Webber, capable of cutting 20,000 board feet of lumber per day, was taken over by William Heywood and S. H. Harmon in 1872. In 1876 Heywood and Harmon, together with Charles L. Dingley and William Bikler, transferred their holdings to a new firm, the Gualala Mill Company. In 1889 the company replaced the mill's circular saws with band saws. The Gualala Mill Company came to an end in

1903, when a new firm, the Empire Redwood Company, took control. Empire Redwood was composed of millmen who had come to the Pacific Coast from the East, and the new concern paid $650,000 for Gualala Mill's properties, by then made up of a large sawmill, a railroad, shipping facilities, and 24,000 acres of timberland. In September 1906, fire totally destroyed the mill. The loss, estimated at over $200,000, was covered by insurance only up to sixty percent. In the 1920's, the mill was no longer operating.

Little River Mill Company, 1864: Formed by Silas Coombs, Ruel Stickney, and Tapping Reeves, who built a mill at Little River that had a capacity of 20,000 board feet of lumber per day. One of the first double circular mills in the county, it employed 100 men. In 1871 Reeves sold his interest to his two partners, and in 1872 Stickney sold his interest to Charles Perkins. The mill burned in March 1874, but the Little River Mill Company, which at that time owned 1,800 acres of timberland, rebuilt it. In April 1878, the company purchased the Murray Mill at Elk Creek, and then moved it to Stillwell Gulch, a mile and a half south of Little River, where it soon employed fifty men, who could turn out 20,000 board feet of lumber per day. The mill at Little River continued to operate at peak capacity until the Depression of 1893 closed it down for good. Its machinery was sold, and in 1910 the mill itself burned to the ground.

The bull donkey crew of the Little River Mill Company pose in the woods for the photographer. — CLARKE MUSEUM COLLECTION

Garcia Mill, 1870: Built by Stevens and Whitmore on the Garcia River and Brush Creek, five miles east of Point Arena. This mill could cut 40,000 board feet of lumber per day and totaled eight million board feet in one eight-month season. In 1872 the property was sold to R. Nickerson and S. Baker. Lumber at the mill moved directly from the saw or planer into a box flume two and one-half feet wide. Four miles away, at the other end of the flume, a water wheel turned a number of rollers that hoisted the lumber 200 feet to the top of the river bank, where it was loaded onto cars and carried two and one-half miles to its shipping point over the Garcia and Point Arena Railroad. L. E. White purchased the mill in 1891, and in 1894 it became part of the L. E. White Lumber Company. That same year, the mill, hotel, and other nearby buildings burned. L. E. White Lumber did not rebuild, because the market was dull and its new mill at Greenwood was easily filling all the company's orders.

L. E. White Lumber Company, 1872: Lorenzo E. White, originally from Massachusetts, built the Whitesboro mill on Salmon Creek and installed the Salmon Creek Railroad about 1876. By 1890 all the timber had been cut, and the mill closed. Earlier, in the mid-1880's, White had acquired timberlands near Greenwood Creek (ten miles from Salmon Creek), and he now moved the Whitesboro mill to Greenwood and put it into operation there. He also bought the old Helmke mill, about two and one-half miles up Greenwood Creek. To acquire more storage space for his logs, he built a dam across the creek and, as a result, the supply of timber waiting to be cut increased to the point where the mill could keep running all year round. In 1890 the mill was producing 80,000 board feet of lumber each day. In 1891 White took control of the Garcia Mill and 7,000 acres of timberland. He built up a thriving lumber business by supplying rail ties to the Southern Pacific Railroad. The ties were manufactured at the Garcia Mill and flumed to the shipping point near Point Arena. By 1894 all the timber in the area to which White had moved the Whitesboro Mill was also logged out, and the mill then closed down for good. L. E. White died two years later, and in 1896 his heirs formed the L. E. White Lumber Company, which was sold to the Goodyear Redwood Company in 1916 for some $3.5 million. The transaction included 85,000 acres of timberlands, a town, both the Whitesboro and Garcia mills, and a line of steamers. C. A. Goodyear and James A. Lacey of Chicago and James Mackenzie of San Francisco comprised Goodyear Redwood. Its parent organization was the C. A. Goodyear Lumber Company,

The L.E. White Lumber Company at Elk which was sold to the Goodyear Redwood Company in 1916. — NATIONAL MARITIME MUSEUM — SAN FRANCISCO

The Goodyear Redwood Company at Elk in 1916. The Depression closed the mill down in 1932. — NATIONAL MARITIME MUSEUM — SAN FRANCISCO

in turn associated with the Great Southern Lumber Company of Louisiana. C. A. Goodyear Lumber also owned property in Washington and 25,000 acres of pine in Glenn County, California. Goodyear Redwood, however, fell victim to the Depression. It went broke and ceased to function in 1932. Subsequently, the Elk River Company took over the Greenwood Creek operation. In 1936 the mill there, which was uninsured, burned down—an event that marked the end of the redwood industry in that area.

The Noyo mill. In 1891 the Noyo interests merged with Fort Bragg Redwood to form the Union Lumber Company. — NANNIE M. ESCOLA COLLECTION

The Cottoneva Lumber Company, 1877: W. R. Miller, who had already built a wire chute at Cottoneva Creek, also erected a double circular mill there, capable of producing 25,000 board feet of lumber per day. In 1887 the Cottoneva Lumber Company, with Joseph Viles of Santa Rosa as its major stockholder, was formed and took over the Miller development in the Cottoneva Creek area, then known as Rockport. Plans to rebuild the mill after a fire came to halt in 1900 because of disputes over ownership of timberlands, and in 1907 the New York & Pennsylvania Lumber Company purchased the Cottoneva Mill, as well as the Hooper Mill on Hardy Creek. Soon afterward, Cottoneva Lumber regained its holdings, but in 1912 a second fire destroyed the mill completely. Between 1924 and 1926, Finkbine-Guild, a Mississippi firm, built a new electric mill at Rockport and modernized the surrounding town. The Southern Redwood Corporation, however, took over the property in 1928 and closed down the new mill the following year. Finkbine-Guild subsequently recovered the mill from Southern Redwood, but in 1931 the Iowa-Des Moines National Bank foreclosed on Finkbine-Guild. In 1938 Ralph Rounds of Wichita, Kansas, a partner in Rounds &

Porter & Associates, retail lumber dealers in Kansas and Oklahoma, purchased the Rockport holdings and established the Rockport Redwood Company, of which Rounds was president and Frank Kilpatrick general manager. Finished lumber was taken to Fort Bragg to be shipped out of Mendocino County by rail. In September 1942, the Rockport Mill burned down, but Rockport Redwood had a new one operating by the following July. In February 1948, the company constructed a seasoning yard and finishing plant just north of Cloverdale, which was known as Rounds & Kilpatrick. Rockport Redwood closed its sawmill in 1957, and in 1967, Georgia-Pacific bought out the Rounds industries. The facilities at Cloverdale were taken over by the new Louisiana-Pacific Corporation in 1973.

Union Lumber Company, 1878: Calvin Stewart and his brother-in-law, James Hunter, began building a mill at Newport in 1876. In 1883 C. R. Johnson purchased an interest in the business, and the firm's name became Stewart, Hunter & Johnson. The Newport Mill proved too small for the firm's plans, and its landing point was a poor one, so in 1884 C. R. Johnson chose the former army post at Fort Bragg as a suitable location for a mill, and the partners formed the Fort Bragg Redwood Company, which started constructing the new facility. Fort Bragg Redwood eventually became one of the three largest redwood projects in the world. In 1891 W. P. Plummer and C. L. White merged their timber interests on the Noyo River with those of Fort Bragg Redwood, to form the Union Lumber Company. In succeeding years, Union Lumber continued to expand by acquiring companies and timber and increased production by improving the mill at Fort Bragg. C. R. Johnson kept the firm going through the bleak years of the Depression, and after World War II, the company resumed its expansion activities. On January 15, 1969, Union Lumber merged with the Boise Cascade Corporation of Boise, Idaho, and in 1973 Boise Cascade sold the former Union Lumber redwood mill in Fort Bragg to the Georgia-Pacific Corporation.

De Haven Lumber Company, 1881: Thomas Pollard (a San Francisco shipping agent later associated with E. J. Dodge of the Eel River Valley Lumber Company in Humboldt County) and Blaisdell built a mill at Wages Creek in Ten Mile Township, but in 1882 their enterprise failed and was purchased by Gill, Gordon, and McPhee, who ran the Wages Creek Mill until 1889, when it closed. That same year, experienced operators of several mills in the Wages Creek area that had shut down as a result of financial problems got together at De Haven, at the mouth of Wages Creek, and formed the De Haven Lumber Company. Its directors were David Gill, Alexander Gordon, and John A. Gordon of De Haven and Thomas Pollard and E. J. Dodge of Alameda. The new company found itself unable to survive the depressed markets of 1900 and 1901, however, and the personal property of De Haven Lumber went up for sale by the sheriff. The Pollard Lumber Company, formed by men who had held large interests in De Haven Lumber, took over the defunct firm's property in 1902, and by 1903 the double circular mill at Wages Creek was operating successfully. Pollard Lumber controlled 6,000 acres of timber, and its lumber was loaded by wire chute at Westport at the rate of 15,000 board feet per day. In the 1920's, the mill was no longer operating.

Southern Humboldt Lumber Company, 1882: Not long after he started to build a wharf at Bear Harbor, a small coastal indentation on the Mendocino coast twenty miles north of Fort Bragg, C. C. Milton accidentally drowned at Rockport. The wharf, from which he had planned to ship tanbark and railroad ties, was left unfinished. Late in 1884, W. A. McCormack of Mendocino City brought it to completion. The following summer, McCormack built a chute, and schooners began coming into Bear Harbor to be loaded with tanbark. By 1890 Weller and Stewart had taken control of the tie and bark business in the area. In 1892 Calvin Stewart and James Hunter, former owners of the Fort Bragg Redwood Company, and Thomas Pollard and Edward Dodge, owners of the Eel River Valley Lumber Company in Humboldt County, got together to purchase the Bear Harbor wharf and 12,000 acres of timberland. On July 26, 1893, they incorporated the Bear Harbor Lumber Company with a capital stock of $200,000 and built two miles of railroad. The following year, Bear Harbor Lumber extended its wharf 100 feet and began gradually extending its railroad inland to a terminus at Indian Creek. There the company constructed an engine house, shops, and a warehouse. Lew Moody erected a hotel and saloon nearby, and the area became known as Moody. On September 8, 1896, Bear Harbor Lumber incorporated the Bear Harbor & Eel River Railroad with a capital stock of $200,000, to take over operation of the rail line and extend it beyond Indian Creek in a northerly direction. In 1902 Henry Neff Anderson, A. W. Middleton, and John A. McPherson of the Anderson & Middleton Lumber Company of Aberdeen, Washington joined Pollard, Dodge, and Stewart, and together they incorporated the Southern Humboldt Lumber Com-

pany with a capital stock of $500,000. Southern Humboldt Lumber announced it would build a large sawmill, and Pollard, Dodge, and Stewart subsequently sold their interests in the new firm to the men from Washington, whereupon H. N. Anderson became Southern Humboldt's president. By September 1903, some 160 men were at work extending the Bear Harbor & Eel River Railroad from Moody to the Eel River, a distance of seventeen and one-half miles. The following year, Southern Humboldt Lumber began building a sawmill at its Camp 10, now called Andersonia, across the river from Piercy. The company also built a dam at Andersonia, making it possible to store 20 million board feet of timber, and began constructing a new wharf at Bear Harbor. On October 28, 1905, with the railroad extension completed and final work on the mill under way, disaster struck. A timber brace being pulled by a steam donkey engine hit the back of Neff Anderson's head. He died on November 6, and years of litigation followed. The mill never operated, and in 1921 mill, machinery, and equipment were dismantled and put in storage. Heavy rains during the winter of 1925-1926 caused high water on Indian Creek, and the dam broke, sending logs that had been stored for twenty years hurtling down the Eel River. In 1940 the Anderson estate instituted legal proceedings for recovery of his property, and by 1947 his grandsons had formed the Indian Creek Lumber Company and erected a

An unidentified engineer running the huge donkey in the Southern Humboldt Lumber Company woods. — DAVID SWANLUND

208

new mill on the Andersonia site. This facility operated for several years, but ultimately fell victim to the many problems that beset the lumber industry during World War II and had to be shut down. Early in 1950, however, Tom Dimmick and Anderson's grandsons came to an agreement: Dimmick would lease the Andersonia mill from them and receive cutting rights to their timber. Gradually modernized, the mill then ran for twenty-two years until, in 1972, it ran out of timber. During that period, approximately 20,000 acres were logged, including 6,000 acres that belonged to Dimmick.

Pudding Creek Lumber Company, 1888: Formed by Captain Samuel Blair and his partner, Alex McCollum. Pudding Creek Lumber's mill, located thirty-six miles east of Fort Bragg, remained in operation for ten years, during which period the company used the facilities of the Fort Bragg Redwood Company to ship its product. In 1903 C. R. Johnson of the Union Lumber Company converted Pudding Creek Lumber into a new company, the Glen Blair Redwood Company. Johnson held a one-fourth interest in this new firm, and John Sinclair, former superintendent of the Pacific Lumber Company at Scotia, put the Pudding Creek Mill back into operation. Two years later, in 1905, Glen Blair Redwood merged with the Union Lumber Company.

Usal Redwood Company, 1889: J. H. Wonderly of Grand Rapids, Michigan was president when the company began constructing its mill at Usal, on the northern Mendocino coast. Instead of building a chute for loading, Usal Redwood built a 1,600-foot dock. By 1891 the mill and a three-mile railroad were completed. In 1894 Robert Dollar of the Dollar Lumber Company took control of the Usal Redwood Company. At the time, Dollar Lumber was operating a mill near Guerneville (Sonoma County), but was gradually running out of timber. Eventually, Dollar also got into the shipping business, forming the Dollar Lines. In 1896 his first steamer, the *Newsboy*, was transporting lumber from the Usal Mill to San Francisco. The Usal Redwood Company came to an end in 1902, when fire destroyed its mill, a warehouse, a schoolhouse, and a county bridge. The Mendocino Lumber Company purchased the steel rails from Usal Redwood's railroad for use in rebuilding railroad tracks near the Big River Mill.

Hardy Creek Lumber Company, 1895: Built a mill at Hardy Creek, between Rockport and Westport. In 1898 C. A. Hooper, part owner of the Excelsior Redwood Company at Eureka, took title to Hardy

The Glen Blair Redwood Company in 1890. This plant, which was located just east of Fort Bragg, was later operated as a subsidiary of the Union Lumber Company. — NANNIE M. ESCOLA COLLECTION

The schooner *Rio Rey* loading tanbark at Usal on the northern Mendocino Coast. Instead of a chute, the Usal Redwood Company built a 1,600-foot dock. — NANNIE M. ESCOLA COLLECTION

Creek Lumber, including its mill and all its timberlands, and the Hooper brothers constructed a new mill, which went into operation in 1902 with access to 11,000 acres of timber. Most of the machinery for this new Hardy Creek mill came from the old Excelsior Redwood Mill on Gunther Island on Humboldt Bay. It could cut 40,000 board feet of timber per day, and its finished product was loaded by wire chute and shipped to Los Medanos, where it was seasoned for marketing in the East. In 1907 the New York & Pennsylvania Redwood Company purchased the Hardy Creek Mill. In the 1920's, it was no longer operating.

Northwestern Redwood Company, 1901: Built a mill with a capacity of 40,000 board feet on Willits Creek, two and one-half miles from the town of Willits. This San Francisco-based company's timber grew on the eastern slope of the Coast Range. When its mill burned down, in 1902, Northwestern Redwood rebuilt immediately, and its new mill could cut 65,000 board feet of lumber in ten hours. In 1905 W. A. S. Foster reported that Northwestern Redwood owned 10,000 acres of timber, and that its mill was cutting 95,000 board feet of lumber per day. The Irvine & Muir Lumber Company was always closely associated with Northwestern Redwood. In 1919, when Irvine & Muir Lumber owned a mill at Two Rock and another at Irmulco, it deeded the latter to Northwestern Redwood. In 1928 Irvine & Muir Lumber Company ended its corporate life and was succeeded by the Irvine & Muir Company. That same year, when the Irvine & Muir Company took over Northwestern Redwood, all properties of both earlier companies came under its control.

The Alpine Lumber Company, 1902: Built a mill at Alpine, nineteen miles up the Noyo River, under the direction of A. D. Duffey, the company's president and manager. In the 1920's, the mill was no longer in operation.

Bibliography

Books

Beal, Scoop. *The Carson Mansion*. Eureka: Times Publishing Company, 1973.

Bean, Walter E. *California, An Interpretive History*. 2nd. ed. New York: McGraw-Hill Book Company, 1973.

Bolton, Herbert Eugene, ed. *Fray Juan Crespî: Missionary Explorer on the Pacific Coast, 1769-1774*. Berkeley: University of California Press, 1927.

Carranco, Lynwood and Labbe, John T. *Logging the Redwoods*. Caldwell, Idaho: The Caxton Printers, Ltd., 1975.

Carroll, Charles F. *The Timber Economy of Puritan New England*. Providence: Brown University Press, 1973.

Caughey, John Walton. *California*. 2nd. ed. New Jersey: Prentice Hall, 1965.

Coman, Edwin T. and Gibbs, Helen M. *Time, Tide, and Timber: A Century of Pope and Talbot*. Stanford: Stanford University Press, 1949.

Cox, Thomas R. *Mills and Markets*. Seattle: University of Washington Press, 1974.

Coy, Owen C. *The Humboldt Bay Region, 1850-1875*. Los Angeles: The California State Historical Society, 1929.

Duke, Donald and Kistler, Stan. *Santa Fe . . . Steel Rails Through California*. San Marino: Golden West Books, 1976.

Eddy, J. M. *In the Redwood's Realm*. San Francisco: D. S. Stanley & Co., 1893.

Elliott, Wallace W. *History of Humboldt County, California*. San Francisco: W. W. Elliott & Co., 1881.

Genzoli, Andrew and Martin, Wallace. *Redwood Pioneer, A Frontier Remembered*. Eureka: Schooner Features, 1972.

Gwinn, J. M. *History of the State of California and Biographical Record of Coast Counties, California*. Chicago: The Chapman Publishing Company, 1904.

Hamm, Lillie E., publisher. *History and Business Directory of Humboldt County*. Eureka: Daily Humboldt Standard, 1890.

Holbrook, Stuart H. *Holy Old Mackinaw*. New York: The Macmillan Company, 1957.

Irvine, Leigh H. *History of Humboldt County, California*. Los Angeles: Historic Record Company, 1915.

Jackson, Walter A. *The Doghole Schooners*. Mendocino: Bear & Stebbins, 1977.

Kneiss, Gilbert H. *Redwood Railways*. Berkeley: Howell-North, 1956.

Leeper, David R. *The Argonauts of Forty-Nine*. South Bend: J. B. Stoll & Company, 1894.

McCullough, Walter F. *Woods Words — A Comprehensive Dictionary of Loggers Terms*. Oregon: The Oregon Historical Society and the Champoeg Press, 1958.

McNairn, Jack and MacMullen, Jerry. *Ships of the Redwood Coast*. Stanford: Stanford University Press, 1945.

Melendy, Howard Brett. "Two Men and a Mill." In Lynwood Carranco, ed., *Redwood Country: History, Language, and Folklore*. Dubuque: Kendall/Hunt, 1971.

Menefee, C. A. *Historical and Descriptive Sketch Book of Napa, Sonoma, Lake and Mendocino Counties*. Napa, California, 1873.

Moungoven, Julia and Escola, Nannie. *The Saga of Little River, 1854-1965*. Fort Bragg: Mendocino County Historical Society, 1966.

Palmer, L. L. *A History of Mendocino County*. San Francisco, 1881.

Puter, S. A. D. and Stevens, Horace. *Looters of the Public Domain*. Portland: The Portland Printing House, 1908.

Ryder, David Warren. *Memories of the Mendocino Coast*. San Francisco: Taylor & Taylor, 1948.

Souvenir of Humboldt County. Eureka: Times Publishing Co., 1902.

Documents

Federal:

Annual Report of the United States Commissioner of Patents, 1869.

Dolbeer, John. *Logging Engine*. United States Patent Office, Patent No. 256,553, April 18, 1882.

Dolbeer, John. *Logging Engine*. United States Patent Office, Patent No. 290,755, December 25, 1883.

Dolbeer, John. *Logging Locomotive*. United States Patent Office, Patent No. 290,756, December 25, 1883.

Report of the Commissioner of the General Land Office, 1888.

State:
California State Board of Forestry. *First Biennial Report, 1885-86.*

County:
Humboldt County Clerk. *Articles of Incorporation.*
Humboldt County Recorder. *Deeds.*

Interviews

Allen, George. Former Superintendent of the Holmes-Eureka Lumber Company. Fortuna, California, various dates.

Barnum, Robert. Head of the Barnum Timber Company. Eureka, California.

Beach, Richard. Former timber faller for various lumber companies. Trinidad, California, various dates.

Bishop, Lois. Public Relations Officer for the former Georgia-Pacific Corporation at Samoa, and now for the Louisiana-Pacific Corporation. Samoa, California, various dates.

Dethlefs, Carl. Head Bookkeeper, McIntosh Lumber Company. Blue Lake, California.

Erickson, Karen. Editor of *Simpson Magazine.* Seattle, Washington.

Fraser, Frank. Former Superintendent of the California Barrel Company's woods. Fortuna, California, various dates.

Fritz, Emanuel. Professor Emeritus of the University of California School of Forestry. Eureka, California.

Hartley, James. Former Public Relations Officer for the Simpson Timber Company at Arcata. Arcata, California, various dates.

Knab, George. Former Administrator for Little River Redwood Company, Hammond Lumber Company, and Arcata Redwood Company. Arcata, California, various dates.

Libbey, Howard. Former President of the Arcata Redwood Company. Eureka, California.

Murphy, Stanwood. Former President and Chairman of the Board of The Pacific Lumber Company. Scotia, California.

Parker, Stanley. Administrator for The Pacific Lumber Company. Scotia, California.

Peterson, Arthur. Former Production Superintendent for the Georgia-Pacific Corporation. and the Louisiana-Pacific Corporation at the Samoa Division. Samoa, California, various dates.

Rutledge, Peter. Former Superintendent for the Dolbeer & Carson Lumber Company. Eureka, California, various dates.

Ryan, Harry. Former woodsman for various lumber companies. Samoa, California, various dates.

Schmitt, Dr. Stanwood. Eureka physician and logging buff. Eureka, California.

Schroeder, Darrell. President of the Miller Redwood Company. Crescent City, California, various dates.

Sundquist, Louis. Former engineer for the Arcata and Mad River Railroad. Blue Lake, California, various dates.

Trobitz, Henry K. California Resources Manager for Simpson Timber Company. Arcata, California.

Vaughn, B. J. Public Relations Officer for Boise Cascade, Union Lumber Region. Fort Bragg, California.

Wallace, Glenn. Simonson Lumber Company. Smith River, California.

Maps

1865 Doolittle Map of Humboldt County

Newspapers

Arcata Union.
Blue Lake Advocate.
Fort Bragg Advocate.
Eureka Humboldt Standard.
Eureka Humboldt Times.
Eureka Times-Standard.
Eureka West Coast Signal.
San Francisco Alta California.
San Francisco Bulletin.
San Francisco Chronicle.

Pamphlets

Background on Fairhaven Pulp Mill. Crown Simpson Pulp Company.
Oral History Interview with P. J. Rutledge. St. Paul, Minn.: Forest History Foundation, 1953.
Our Redwood Heritage. California Redwood Association.
Welcome to Fairhaven. Crown Simpson Pulp Company.
Welcome to Hammond Lumber Company. Hammond Lumber Company.
Welcome to the Sawmill. Georgia-Pacific Corporation.

Periodicals

Borden, Stanley T., "The Albion Branch." *The Western Railroader* XXIV (December 1961): 3-30.

— ."Bear Harbor & Eel River Railroad." *The Western Railroader* XXVII (May 1964): 1-8.

— ."Caspar Lumber Company." *The Western Railroader* 315-316 (1966): 3-33.

Bronson, William. "Behind the Redwood Curtain." *Cry California* I-4 (Fall 1966): 10-20.

Carranco, Lynwood. "Logger Language in the Redwood Country." *Journal of Forest History* 18-3 (July 1974): 52-59.

Carranco, Lynwood and Fountain, Mrs. Eugene. "California's First Railroad: The Union Plank Walk, Rail Track, and Wharf Company Railroad." *Journal of the West* III (April 1964): 243-254.

Erickson, Karen. "Expanded Redwood Park Bill Is Signed." *Simpson Magazine* (April 1978): 9-10.

— ."Reforestation." *Simpson Magazine* (January 1978): 4.

— ."Test-Tube Forest with Bull-Team Roots." *Simpson Magazine* (April 1978): 2-5.

Grosvenor, Melville Bell. "World's Tallest Tree Discovered." *National Geographic* (July 1964): 1-9.

Kortum, Karl and Olmsted, Roger. "A Dangerous-Looking Place — Sailing Days on the Redwood Coast." *California Historical Quarterly* XLX (March 1971): 43-58.

Martin, Wallace. "Tom Kennedy — Donkey Engineer." *Humboldt County Historical Society Newsletter* IX (July 1961): 7-8.

McCloskey, Michael. "The Last Battle of the Redwoods." *The American West* VI (September 1969): 64-70.

Palais, Hyman. "Pioneer Redwood Logging in Humboldt County. *Forest History* 17-4 (January 1974): 18-27.

Palais, Hyman and Roberts, Earl. "The History of the Lumber Industry in Humboldt County." *Pacific Historical Review* XIX (February 1950): 1-11.

Redwood Log. Samoa, California: Hammond Lumber Company, 1948-1951.

Wood and Iron. San Francisco, California, 1884-1909.

Zahl, Paul A. "Finding the Mt. Everest of All Living Things." *National Geographic* (July 1964): 10-51.

Proceedings

Redwood Region Logging Conference. *Proceedings of the 1960 Logging Conference*. Eureka, California, 1961.

Unpublished Materials

Melendy, Howard Brett. "One Hundred Years of the Redwood Lumber Industry, 1850-1950." Ph.D. dissertation, Stanford University, 1952.

Mengel, Lowell S. "A History of the Samoa Division of Louisiana-Pacific Corporation and Its Predecessors, 1853-1973." Master's thesis, Humboldt State University, 1974.

Wattenburger, Ralph T. "The Redwood Lumber Industry on the Northern California Coast, 1850-1900." Master's thesis, University of California, 1935.

Wood, Claudia. "The History of the Pacific Lumber Company.," Term paper, Humboldt State College, 1956.

Index

(* Asterisk denotes illustration)